FORTRESS E[illegible]E

THE ATLANTIC WALL GUNS

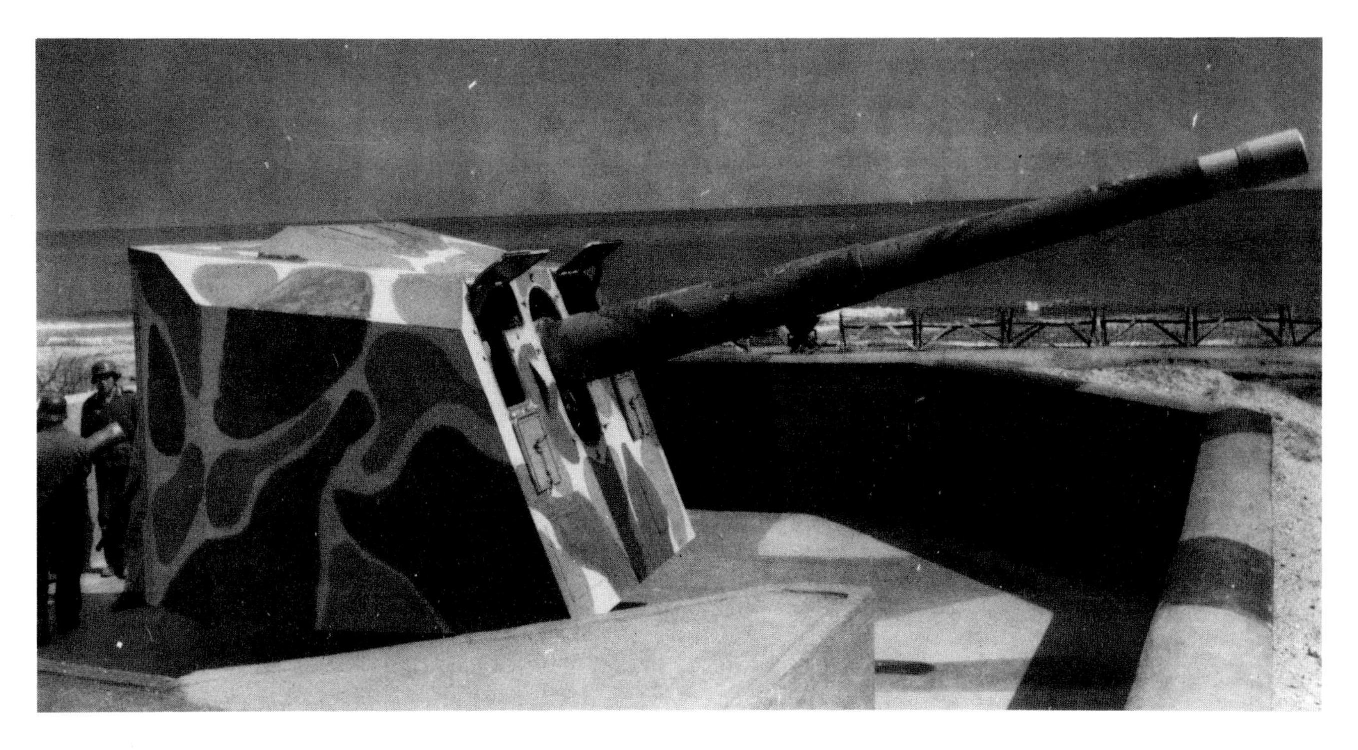

A 15 cm TbtsK C/36 of Battery M II in an open firing position. At the left edge of the picture, the loader can be seen with a cartridge.(BA)

KARL-HEINZ SCHMEELKE & MICHAEL SCHMEELKE

Schiffer Military/Aviation History
Atglen, PA

SOURCES

Federal Archives – Military Archives, Freiburg
Delefosse, Atlantikwall-Heer, Abbeville 1988
Delefosse, La Batterie Todt, Abbeville 1988
Gamelin, Objectives Douvres, Nantes 1975
Harnier, Artillerie im Küstenkampf, Munich
Perret, Churchill Tank, Osprey Publ. 1980
Ramsey, After the Battle, No. 29, London 1980
Leitfaden der Marineartillerie, Vol.2, Berlin 1940
Uniformen der Marine und Schutztruppen, Munich 1930
Drawings, Chantal and Yannick Delefosse
Photos, Federal Archives, Koblenz 27 (BA)
Imperial War Museum, London 5 (IWM)

ACKNOWLEDGEMENTS

For their kind assistance in the creation of this book, we extend our thanks to Alain Chazette, Chantal and Yannick Delefosse, Heinrich Hartmann, Dr. Jean Pierre van Mol, Meinrad Nielges, Gordon Ramsey, Günter Zietlow, Marton Szigeti and R. Heinz Zimmermann.

Our special thanks are extended to the former members of the Navy: Dr. Theodor Stormer (Marinekorps Flandern), Hans Eckert, Hans Donde, Hans Gebhard, K. H. Riecken (MAA 240), Walter Benz, Reimer Eilers, Willi Schüler (MAA 242), Lothar Fischer (MAA 244), as well as Mr. Werner Zeidler (Gun Commander in the Friedrich August Battalion).

COVER PHOTO
A 40.6 cm SK C/34 in a bunker.

A 40.6 cm gun of the Schleswig-Holstein Battery near Sangatte, The 40.6 cm ship cannon (SK) was the largest caliber used in naval artillery. After the battleship Bismarck was sunk, the battery was renamed in honor of the ship's captain, Kapitän zur See Ernst Lindemann. (BA)

Translated from the German by Edward Force.

This book originally appeared under the title,
Fernkampfgeschuutze am Kanal,
by Podzun-Pallas Verlag, Friedberg.

Printed in the United States of America.
ISBN: 0-88740-525-8

We are interested in hearing from authors with book ideas on related topics.

Published by Schiffer Publishing Ltd.
77 Lower Valley Road
Atglen, PA 19310
Please write for a free catalog.
This book may be purchased from the publisher.
Please include $2.95 postage.
Try your bookstore first.

THE COAST ARTILLERY

HISTORICAL RETROSPECT

The coast artillery has had a long tradition in German naval history. All the harbors of the Imperial fleet were defended by batteries of two to eight guns in various calibers. The manufacturers of these guns were usually the firms of Krupp and Rheinische Metallwarenfabrik, later Rheinmetall.

The original units of the coast artillery were the Matrosenartillerie Units I-V, which were responsible for replacing and training artillerymen as well as for supplying them with guns and equipment.

They were directly subordinate to the Inspector of the coast Artillery and Mines, with headquarters in Cuxhaven.

After the outbreak of World War I, naval batteries were set up hastily to secure the occupied coasts. The strongest concentration of guns took place on the Flemish coast from Knokke to Westende, as an extension of the northern section of the western front.

After the Marinenkorps Flandern was founded in November 1914, with a submarine base in Bruge and torpedo-boat bases and naval air bases in Ostende and Zeebrugge, the first battery of four 5 cm ship cannons was installed on the coast near Duinbergen. This was soon followed by others.

The heaviest gun used by the Imperial Coast Artillery was the 38 cm ship cannon (SK) with a barrel length of L 45 (barrel length + caliber x 45).

This gun, developed by Krupp, was originally planned for the battleships of the Bayern class. While these two ships, the "Bayern" and the "Baden", were launched in 1915, the "Sachsen" and the "Württemberg" were not completed. Eight 38 cm guns in four double turrets had been planned for each of these ships.

The extra barrels were mounted in special firing platforms known as "Bettungsschiessgerüste", and sent to the Pommern battery near Moere and the Deutschland battery near Breedene to be used in coastal defense.

There were two different shells for the 38 cm SK L/45 gun. The 400-kilogram L/4.1 (length = 4.1 x caliber) explosive shell with base fuse plus nose fuse with ballistic sheath had an initial velocity of 1040 meters per second and a maximum range of 47,500 meters.

The 750-kilogram L/5.4 explosive shell with base fuse plus nose fuse with sheath could be fired 38,700 meters with an initial velocity of 890 meters per second. These two batteries fired not only on sea targets, but also on the city and harbor of Dunkerque, where most goods and supplies for the western front were unloaded during World War I.

One gun of the Moere battery remained in place until the fifties and was scrapped only when a modern swimming facility was built.

In addition to the conventional means of aiming, such as the distance clock, long base, and plane-table, ranges were measured and fire directed by means of aerial observation as early as 1915. By 1918 this process had been developed further, and satisfactorily, through the use of radio by the naval land and sea flight units.

A 38 cm SK L/45 gun of the Pommern Battery.

The planned landing operations of the western powers on the Flemish coast with the goal of turning the right wing of the German western front were given up, since the coast artillery had a frightening effect.

The light batteries had already proved themselves during the British commando operations against Ostende and Zeebrugge.

After the end of World War I, the naval artillery units were disbanded. In their place, coastal defense units were established in the Reich Navy, later to be renamed naval artillery units. Since Germany had to surrender all modern guns, the artillery more or less retrogressed to its 1905 state.

In the 1920s the navy tried to modernize its outmoded guns to equal those of foreign navies.

Only after 1933 were major efforts made by the Reichsmarine and Kriegsmarine (after May 21, 1936) to build up coast artillery.

COAST ARTILLERY IN THE WEHRMACHT

With the introduction of universal military training, new men could be trained regularly for the seven naval artillery units at the Naval Artillery School in Swinemünde.

The men of the naval artillery wore a field-gray uniform that differed from that of the army in the golden yellow color of its insignia and braid. The career emblem was a winged shell.

Now new guns, intended primarily to be artillery for ships, were being developed and built by the industry.

In the development of guns for the Type F and G battleships ("Bismarck" and "Tirpitz"), the Krupp engineers remembered the well-developed design of the 38 cm SK L/45 gun of World War I. The barrel length was extended to fifty caliber lengths, and with the rifling to the right increasing from 1/38 to 1/30, the number of riflings was reduced from 100 to 90. By the end of 1934, this further development could be concluded and the gun designated 38 cm SK C/34.

For the ships to be built subsequently to the "Bismarck" and "Tirpitz", the Type H super-battleships, the caliber was increased by Krupp to 40.6 cm. After plans for this type of ship had been canceled, all the 40.6 cm SK C/34 guns were mounted in C/38 bedded firing platforms for use in coastal defense.

When World War II broke out, several of the navy's heavy coastal batteries were moved forward into already prepared positions in the western wall.

The Oldenburg Battery of 24 cm SK L/50 and the Friedrich August Battery of 30.5 cm SK L/50 guns were transferred to the southern Black Forest. From these positions they supported the Seventh Army's attack on the Maginot Line between Strasbourg and Mulhouse, beginning on June 15, 1940.

After the armistice with France, the German Wehrmacht had occupied the Atlantic Coast from North Cape to the Pyrenees.

When Britain, their last opponent in the west, rejected Germany's offer of piece, the invasion of the British Isles was planned under the code name of "Sealion."

The main landing was to be carried out by the 9th and 16th Armies, crossing the channel at its narrowest point from France to the south coast of England between Ramsgate and Brighton.

The Germans believed they could keep the superior British fleet away from the Channel by using long-range artillery. For this purpose, Naval Artillery Units 240, 242 and 244 were formed and stationed between Calais and Boulogne.

These new artillery units were assigned batteries by the Naval High Command (OKM) as follows:

MAA 240

Battery	Friedrich August	3 x 30.5 cm SK L/50
Battery	Créche I	4 x 19.4 cm 486f (captured French battery)

MAA 242

Battery	Siegfried, later Todt	4 x 38 cm SK C/34
Battery	Grosser Kurfürst	4 x 28 cm SK L/50
Battery	M III	4 x 17 cm SK L/40
	(Army battery to 1942, rearmed as naval battery with 15 cm C/36 torpedo-boat guns)	
Battery	M IV	3 x 17 cm SK L/40

MAA 244

Battery	Schleswig Holstein (later Lindemann)	3 x 40.6 cm SK C/34
Battery	Bastion II	3 x 19.4 cm 486f (captured French battery)
Battery	Oldenburg	2 x 24 cm SK L/50
Battery	M I	4 x 17 cm SK L/40
	(Army battery to 1942, rearmed as naval battery with 15 cm SK C/28)	
Battery	M II	4 x 17 cm SK L/40

A gun of the Lindemann Battery at its greatest barrel elevation of 52 degrees. Three of the ten barrels of the 40.6 cm SK C/34 that were produced were used by the Lindemann Battery. (BA)

Sketch map showing long-range batteries on the Channel.

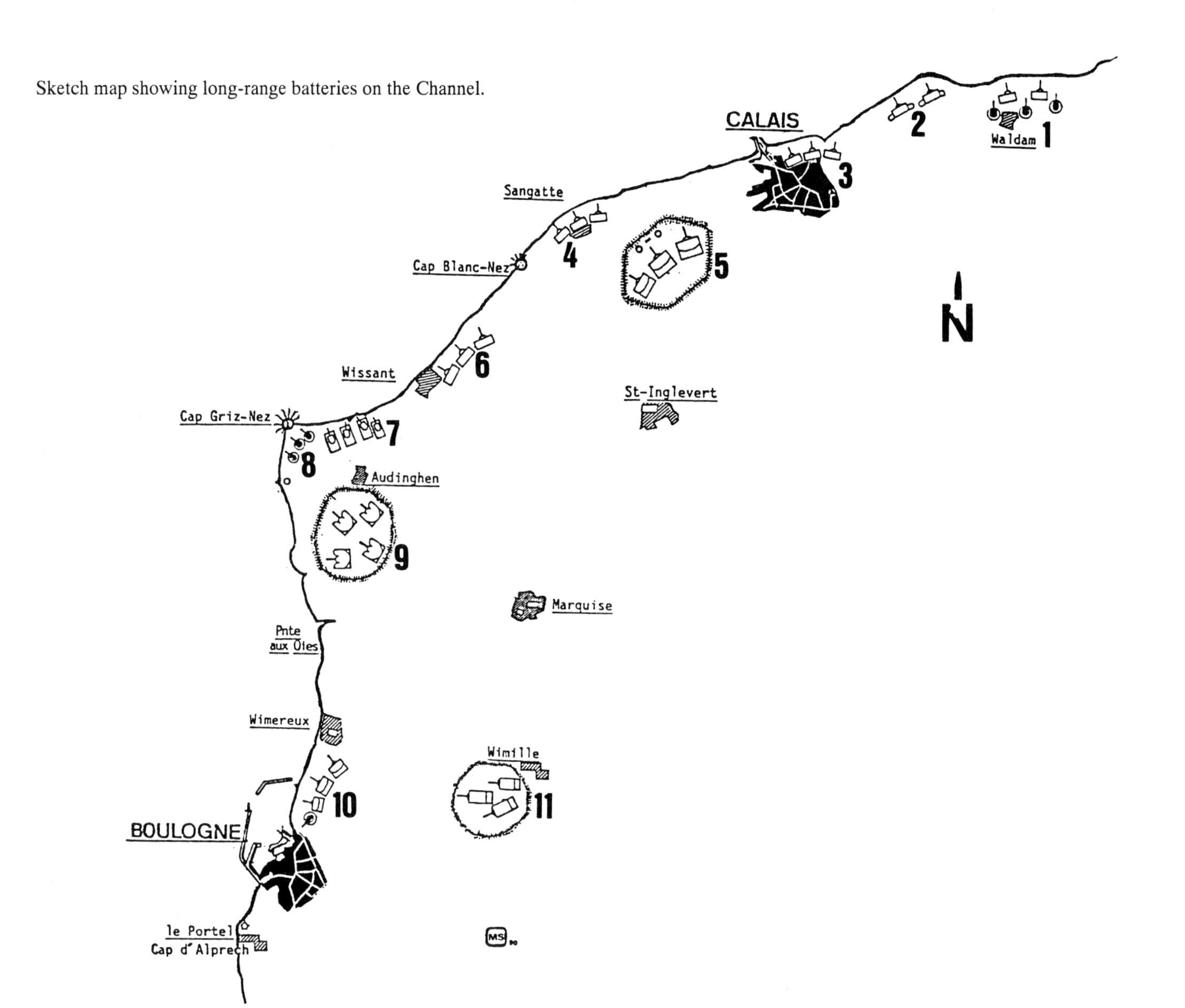

A 17 cm SK L/40 gun of Battery M II near Sangatte, seen in an open firing position in the summer of 1940. For shrapnel protection, a wall of sandbags has been erected around the position.

The batteries were supplied with the most modern fire-control devices available to the navy, from the ten-meter range finder to the radio measuring devices (radar) of the "Freya" and "Würzburg" types. As of April 1941, each unit was also issued several infrared heat-sensitive devices.

In the ensuing months, the coastal area between Calais and Boulogne developed into a single huge construction site. Gun mounts were cast by the Todt Organization and the naval fortress construction staffs with the help of hired French workers, bunkers were built and supply bases set up.

In June of 1940 the Créche I and Bastion II batteries were ready for action. On July 7 the Créche I battery sank a British torpedo boat off Wissant and was duly mentioned in Wehrmacht reports. The third gun, under the command of Feldwebel Hans Gebhard, performed particularly well in this action.

By the end of 1940, all the batteries except the Lindemann were ready for action. The construction work, though, especially that on the gun bunkers, continued into the spring of 1941. The bunkers, with their four-meter-thick walls, protected the crews and their guns from bombs and shells, but they limited the lateral range of fire to about 120 degrees.

The Lindemann Battery was transferred to the Channel coast only in 1941, before which it had been tested for firing technology at Hela.

In addition to the naval batteries, the army also had a number of heavy gun batteries stationed along the channel, including eight railroad batteries with 17 cm to 28 cm caliber guns. The majority of the guns were withdrawn after the Russian campaign began.

In the autumn of 1941, plans to invade Britain were finally given up. The task of the naval batteries on the channel was now limited to long-range firing on England, which meant firing on the south coast of England, including the harbors of Ramsgate, Dover and Folkstone. In addition, the Channel was supposed to be closed to enemy shipping.

From November 1942 to June 1944, the Grosser Kurfürst Battery alone fired more than four hundred shells at the English south coast. In the first fifteen months, all the batteries fired a total of 2450 rounds, 1242 of them at enemy ships.

With the construction of the Atlantic wall, beginning in 1942, the naval artillery batteries on the Channel coast became the nucleus of this new system of fortifications.

It was presumed by the Wehrmacht High Command (OKW) that an Allied invasion could take place only at the narrowest part of the Channel between Britain and occupied Europe.

A gun mount for a 38 cm SK C/34 of the Todt Battery. In the background is a 175-ton Demag crane, with the help of which the gun and armored turret were mounted.

Left: A 19.4 cm gun of the Creche I Battery near Boulogne after being captured by German troops.

All the batteries were now expanded into self-contained defensive facilities, with extended mine fields, electric fences and foglaying devices. The Grosser Kurfürst Battery was even given a wooden mockup of a battery with electric muzzle flash.

The Allied invasion took place in Normandy on June 6. At the beginning of August, the landing troops broke out of their bridgehead, and on August 25 Paris was liberated and the Seine was crossed.

Along the Channel coast, troops of the First Canadian Army stormed ahead. On September 10, the naval artillery units were surrounded in a pocket that extended from Boulogne to Calais.

At the Führer's command, the defensive zone was declared a fortress and was to be held until new offensive armies could be established.

The Todt, Grosser Kurfürst and Lindemann Batteries carefully fired on the English coast, so as to use up their large supplies of ammunition.

On September 13, the Canadians tried, with the 7th Infantry Brigade and tanks, to break into the fortifications at Cape Gris Nez, but were fought off by the naval artillerymen.

Now the OKM's decision to overbunker the guns came back to haunt them. On account of their limited traverse range of 120 degrees and their firing direction toward the sea, only the Grosser Kurfürst Battery, one gun of the Créche I Battery and two guns of the M I Battery could take part in the land battle.

Whether any of the guns could have been taken out of the bunkers and provisionally mounted on the land front is a question that the authors have not been able to answer to date. Information from readers, particularly from former members of the artillery units, would be most welcome.

After heavy bombing attacks, the Canadians were able to break into the fortress at Boulogne on September 17. After heavy fighting for every battery, the Friedrich August and Créche Batteries fell on September 18.

On September 21 all of Boulogne was in Canadian hands. Some of the artillerymen were able to fight their way to Gris Nez and strengthen the defenses there.

Here the turret commander of Caesar Turret, Kapitänleutnant Peschel, examines the bunker for damage after a bomb attack.

Below: The entrance to the Friedrich August Battery is secured with barriers of armor.

On September 25 the Canadians attacked the Lindemann Battery with infantry and three tank units of the 79th Armoured Division.

A series of special vehicles had been built in Britain for the storming of the Atlantic Wall. Mortars and mine launchers in 29 cm caliber had been mounted on the chassis of Churchill Mark III and Mark IV tanks. Their shells, especially loaded with an explosive charge developed to break concrete fortifications, could be fired at a maximum range of 120 meters.

A flamethrower had been mounted on the chassis of the Churchill Mark VII tank. The range of this gun was some 70 to 100 meters. The 400 liters of combustible liquid were carried in an armored trailer. Minesweeping and bridgelaying tanks were also built.

Three units of the 79th Armoured Division were equipped with these special vehicles and thoroughly prepared for the storming of the long-range batteries along the Channel.

Advance bombing attacks on the Lindemann Battery had destroyed almost all the close-range defensive facilities, so that tanks and infantry could push forward quickly to the gun bunkers.

In addition, fire from the Grosser Kurfürst Battery could no longer be of help. All three turrets were shot down by the evening of September 26, particularly through the terrifying effect of the flamethrowing tanks.

The M II Battery at Sangatte fell on the same day. Now the tanks could advance unhindered along the coast road toward Gris Nez and Calais.

On September 29 the final battle for the batteries at Gris Nez and the Todt and Grosser Kurfürst Batteries began.

The crew of Turret I of the Todt Battery surrendered around midday, after a flamethrowing tank had been able to set fire to the interior of the turret by firing through a loophole. Shortly thereafter, Turret IV was also set afire and conquered.

During the afternoon, white flags flew at the command post and the M IV Battery at the cape. The crews of turrets II and III were still able to defend themselves against the Canadians until evening, then they too were overcome.

A few artillerymen were able to escape through the shell craters during the night and avoid being taken prisoner. But only a few, including Feldwebel Benz, were able to reach their homeland.

On September 29 the Grosser Kurfürst Battery was also overwhelmed. The last naval batteries on the channel to fall were the Oldenburg and M 1 Batteries near Calais, which capitulated on September 30. With their fall, the Naval Artillery Units 240, 242 and 244 had ceased to exist.

Below: Right after capturing them, Canadian engineers blew up the majority of the German guns. This picture shows a C/36 pivoting mount, and behind it the 15 cm barrel, after the M I Battery was blown up.

GUNS

15 CM SHIP CANNON C/28 IN COASTAL MOUNT C/38

The 15 cm SK C/28 was used as a weapon aboard the Navy's cruisers and destroyers, and as of 1942 also in the coastal defenses.

The gun had a 44-rifling inner barrel with an outer sheath. The vertical wedge, or dropping-block, breech was situated in the heavy breech section, typical of naval guns, which was intended to keep the recoil as moderate as possible.

In coastal defense, the gun was mounted in a coastal mount, or central pivot mount C/36/ This made a traverse arc of 360 degrees and a barrel elevation of 35 degrees possible.

Explosive shells of the L/4.6 type, nose fuse with sheath (Kz), L/4.5 explosive shells with base fuse (BdZmHb), and L/3.8 antitank shells (PzGr) were fired. The last were also used particularly against armored ship targets.

Above: A 15 cm SK C/28 of the M I battery. The frontal armor of the turret, 10 cm thick, is easy to see; the cover of the targeting scope is open. With it the gunner could observe where the shots fell.

The breechblock of the 15 cm SK C/28 with the falling-block breech. The barrel and breech together weighed 9030 kilograms.

All shells were loaded separately from the shell cartridge. This brass cartridge, later made of iron for lack of materials, contained a 14.1 kg charge of propellant powder.

Along the Channel, Battery M I of MAA 244, after taking over an Army position, was supplied with four guns of this type. At first the guns were mounted in open ring beds; only in 1944 did the OT erect two gun bunkers of Type M 270. One gun was left in an open mount as a so-called all-around gun.

For the last gun, the Navy, in cooperation with the Oberfestungspionierstab in France, built the first rotating steel-and-concrete firing position, beginning in April of 1944. This concrete turret was mounted on a roller mount from a turret of the French battleship "Provence." It was made to carry a load of some 800 tons.

The turret was to be turned with the help of an electric motor, but until the work was abandoned in September of 1944, the gun could be turned provisionally by hand.

No reports exist that concern the performance of this turret in combat.

Left: A concrete turret with a 15 cm gun after the war. The barrel points landward, in the direction from which Canadian forces approached the battery. (Photo: Chazette)

Lower left: Part of the roller track on which the turning turret rested.

Below: The shell-case ejector at the rear of the turret.

Gun Bunker Battery M 1 with rotating concrete cupola

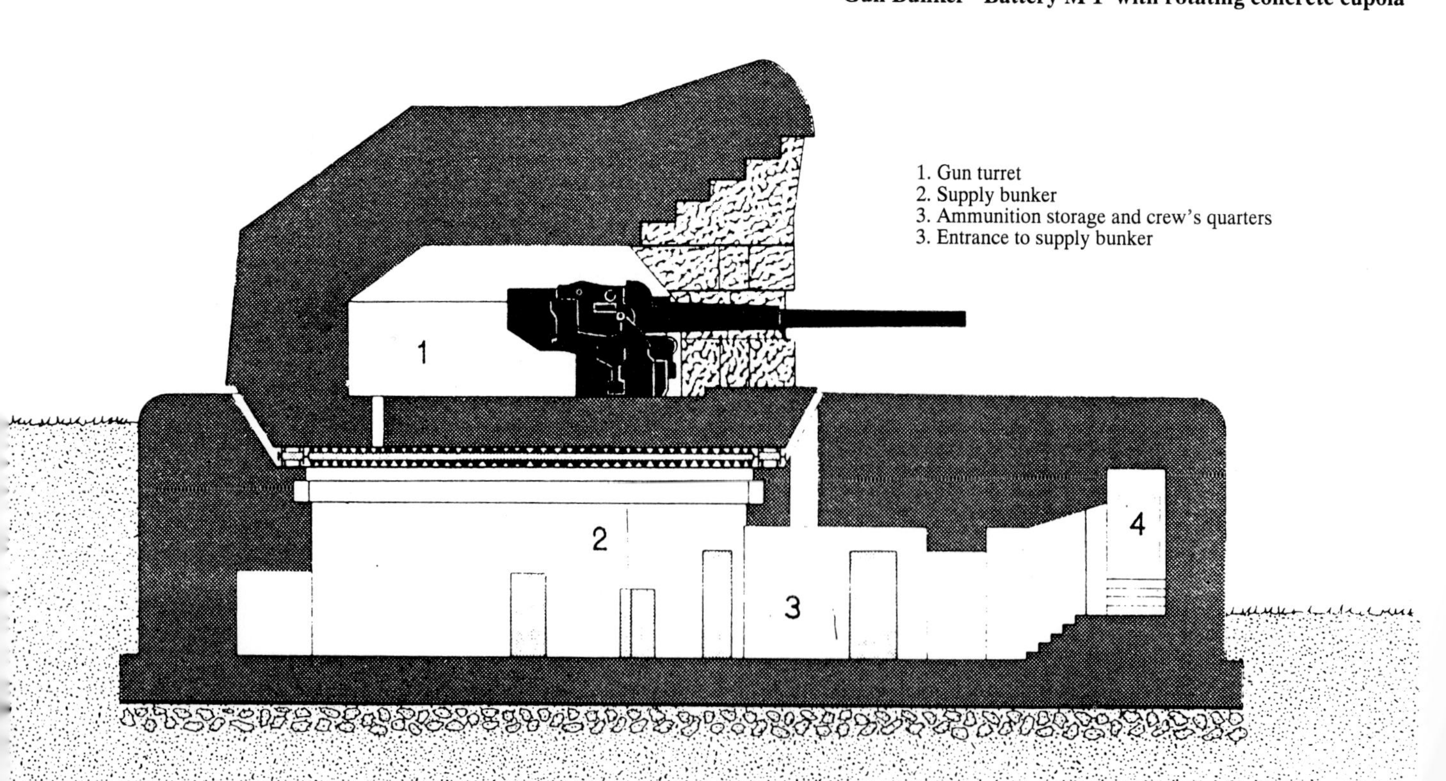

15 CM TORPEDO-BOAT CANNON C/36 (L/45)

The 15 cm torpedo-boat cannon was designed to be essentially identical to the SK C/28. It differed externally from the latter by having a horizontal or wedge breech.

The breech was a slightly wedge-shaped four-cornered block that fit into a horizontal wedge-shaped aperture in such a way that, in a closed position, it pressed firmly against the crimped rim of the cartridge and thus formed a gas-tight closure.

The caliber of the barrel, which had 48 riflings, measured 149 mm, and the maximum shot range was 19.525 meters.

The same types of ammunition were fired from the torpedo-boat cannon as from the SK C/28, and they were likewise ignited electronically.

The Krupp and Skoda firms produced the weapon.

These two photos show the 15 cm C/36 torpedo-boat cannon, made by Skoda, in a bunker. The gun and its crew were very well protected from enemy fire and bombs by the reinforced concrete walls of the bunker, which were two meters thick, but the traverse field was limited to 120 degrees.

Above: A 15 cm TbtsK C/36 of the M II Battery in an open firing position; the gun is still painted gray for use on a ship. (BA)

Below: The same gun at a later point in town. The gun and firing position have now been camouflaged with tan and green paint. A
the left, the K3 loader has just placed a shell in the loading pan; beside him is the telephone by which commands and aiming value
were transmitted. (BA)

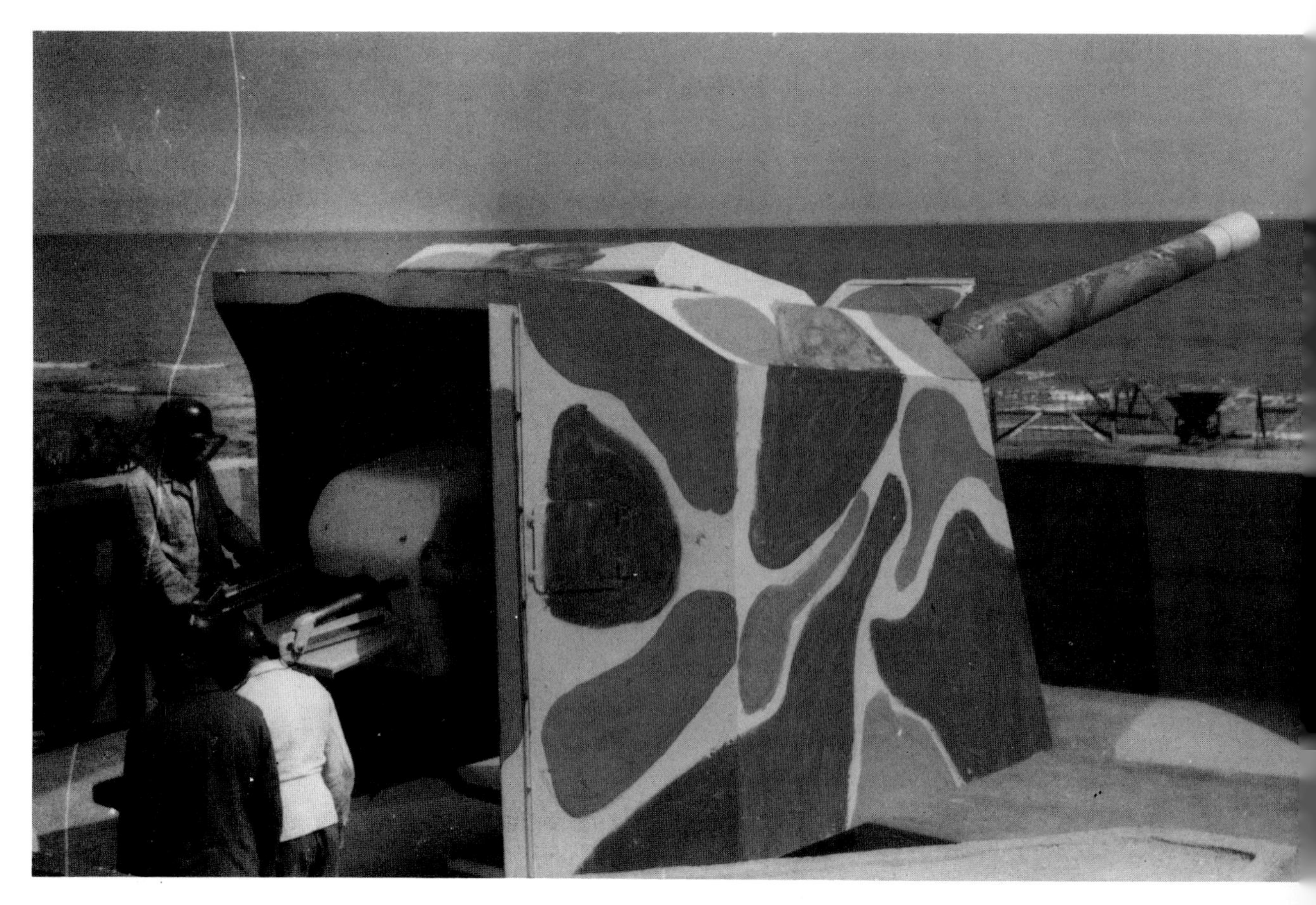

17 CM SHIP CANNON L/40

The 17 cm SK L/40 was one of the oldest guns in the coastal defenses. It had been developed and built since 1902 for the cruisers of the Braunschweig class.

The gun had a horizontal over wedge breech, a barrel length of 4992 mm, a caliber of 173 mm, and a maximum shot range of 26,000 meters.

The inner barrel with its mantle made of several parts rested in a cradle mount. Under the barrel was the hydro-mechanical recoil and recuperator system. This braked the gun's recoil after the shot by means of a brake cylinder that was filled with a mixture of glycerine and water.

In the German Navy it was customary to use a mix of 60% glycerine and 40% water, which enabled the guns to stay ready for action even in low temperatures.

The mechanical recuperator spring was compressed by the recoiling barrel, and then pushed the barrel forward again and into its original position.

As a rule, the gun was set up in open beds or bunkers, with a shield of 10 cm ship's armor plate.

The gun fired the L/4.7 explosive shell with a hood, the antitank shell with base fuse, and the flare shell (LtGes) L/3.4, with a weight of 58.5 kilograms. The flare shell contained an apparatus that, after a time set by the fuse (maximum 60 seconds), was ejected from the base of the shell and lit in the process.

The flare hung on a parachute and burned for about three minutes.

All shells were loaded separately from the cartridge case.

Two pictures of a 17 cm SK L/40 of the M IV Battery at Cape Gris Nez. The lower photo shows a firing drill. The K1 elevation gunner stands by the gun on the right side of the picture, and an ammunition cart can be seen beyond him. To the left is the K2 traverse gunner; a loader stands in front of him with a cartridge.

In the photo at left, the gun has been disguised with reed mats and branches. A wall of sandbags offers shrapnel protection.

19.4 CM CANNON M 70/93 486f (L30.4)

During the French campaign, the German Wehrmacht captured batteries along the French coast. These batteries were armed with M 70/93 flat-fire guns of 19.4 cm caliber.

The M 70/93 gun, produced by the firm of Schneider & Cie. of Le Creusot, were given the German designation of 486f (French) and were included in the coastal defenses.

The basic design dated to 1870 and was built in improved form until 1936.

The gun barrel consisted of a mantle with an inner barrel inside. The length of the barrel was 5550 mm, of which the rifled section was 4600 mm long.

The gun had a longitudinal or screw breech. This was located in the loading orifice and consisted of a cylinder with screw threads on the outside, cut into a part of the circumference.

There were usually three screws and three cut segments, each comprising 60% of the circumference.

Correspondingly, the loading orifice had three smooth segments and three with threadings. The breech was pushed into the loading orifice on a turning swing, turned 60% and bolted in place.

An explosive shell weighing 83 kilograms was fired with the help of a three-part cardboard cartridge. After firing, the barrel had to be cleaned by compressed air to remove residue of the cartridge. This laborious process meant that a shot could be fired only once every 45 seconds.

The bag cartridge was ignited by a friction ignition screw that was screwed into an ignition-pit shaft in the breech.

This ignition apparatus consisted of a brass screw around which a roughened wire was wrapped. The wire stuck out from the screw and could be pulled out with the help of a line. The resulting friction ignited the charge.

For every shot a new friction-ignition screw had to be used.

The maximum shot range was 18,300 meters, with an initial velocity of 638 meters per second.

Above: The Creche I Battery, north of Boulogne, had four 19.4 cm 486f guns. In front is the gun, mounted on an open ring bed, with which the Canadian tanks were fired on. (IWM)

Below: On the lowered barrel of this 19.4 cm gun sits Fw. Gebhard with the battery's mascot, a small brown bear.

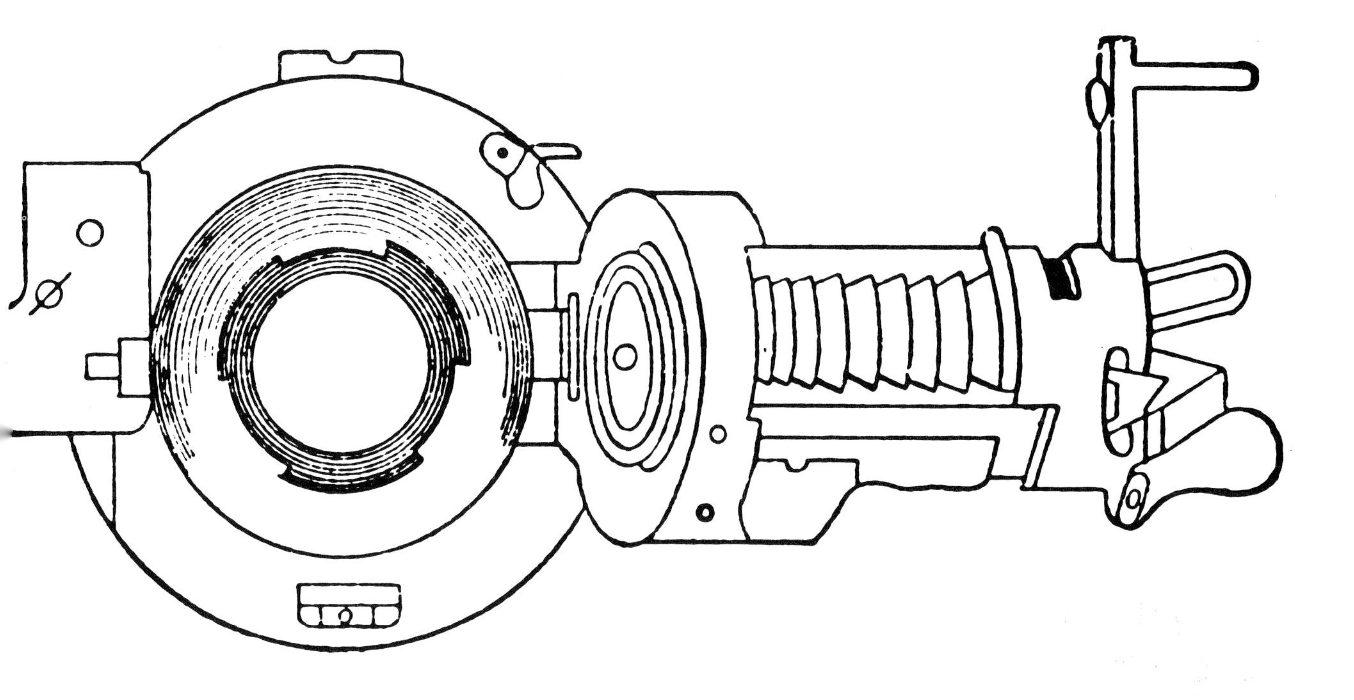

Above: A drawing of the screw breech and loading hole of the 19.4 cm gun. All movements, such as opening and closing the breech, could be carried out with a crank.

Below: The massive screw breech of the 19.4 cm gun when opened.

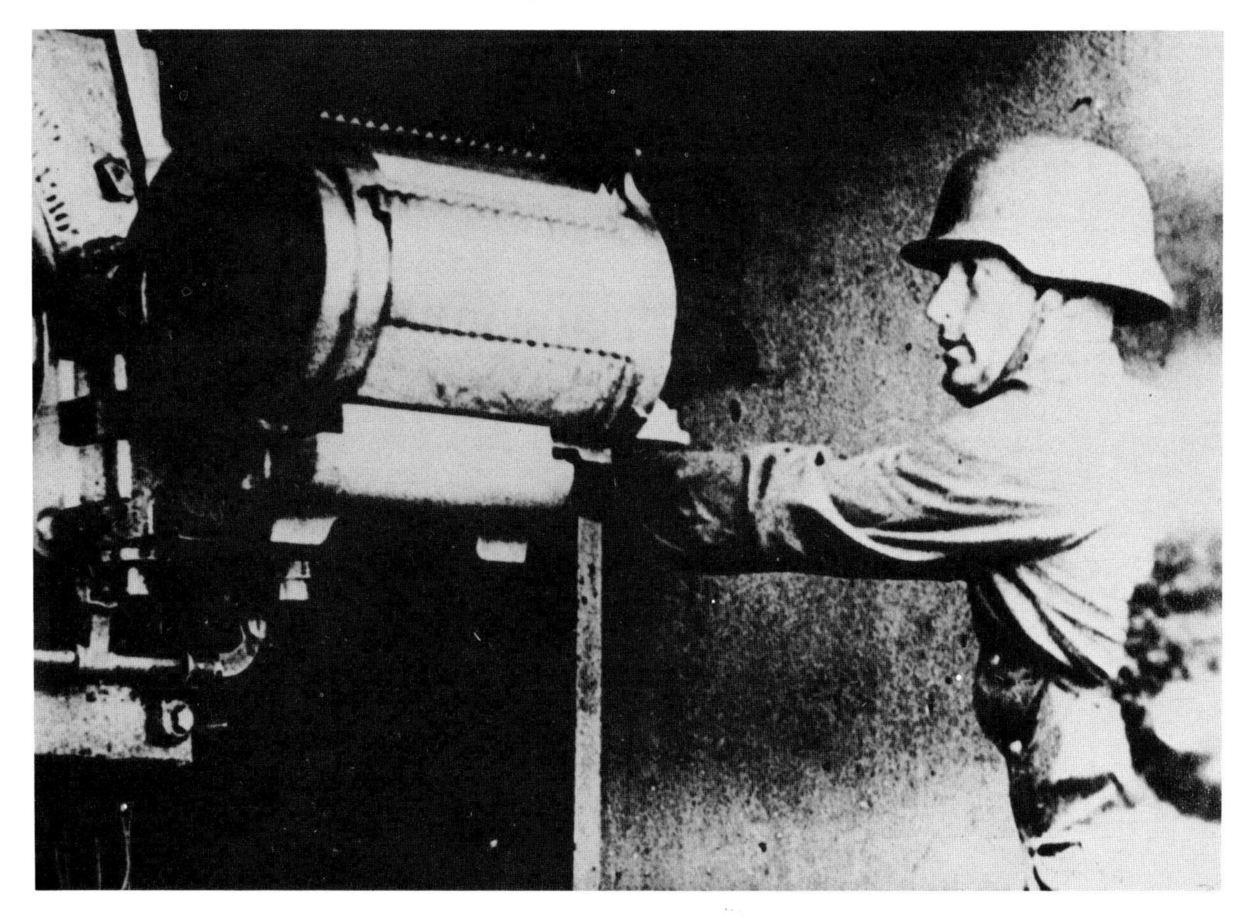

24 CM SHIP CANNON L/50

The two 24 cm guns of the Oldenburg Battery were manufactured in Russia and had been captured by German troops near Libau in September of 1915.

The guns were immediately removed from their beds and sent to Germany for testing. Their caliber was originally 254 mm. In the thirties, Krupp modernized the two guns by inserting new inner barrels of 238 mm caliber into the mantles. Thus the two guns became the equivalent of the German 24 cm type.

The barrel length was 11,900 mm, and the screw breech was retained.

The old mounts with their armored shields allowed only a maximum barrel elevation of 30 degrees. The maximum shot range after 270 shots had been fired remained 28,000 meters, with an initial velocity of 900 meters per second.

An L/4.2 explosive shell with base fuse and nose fuse with sheath, weighing some 148.5 kg, was fired, as was an explosive L/4.1 armor-piercing shell with a wright of 150.5 kg.

Captured Russian guns were often used in the German coastal defenses. They were simply but sturdily built.

After the war, a special mixture of gases had to be used to cut up the Russian 24 cm guns. It was much more expensive than the guns' value as scrap metal.

Left: Infantrymen of the coastal defenses are photographed looking at the 24 cm gun of the Oldenburg Battery in the dunes near Calais. (BA)

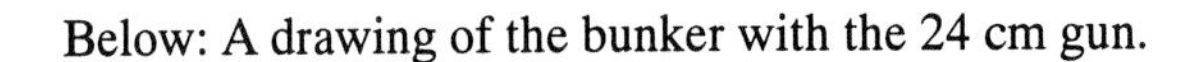

Below: A drawing of the bunker with the 24 cm gun.

Above: The Oldenburg Battery's two gun bunkers, with the artillerymen's barracks behind them.

Below: A 24 cm SK L/50 barrel in a raised firing position. The old mounts allowed a maximum elevation of only 30 degrees. (BA)

Below: This picture shows the mount and the 10 cm protective armor plate. The barrel is in loading position, +2 degrees. (BA)

A PZ drawing of the Oldenburg Battery's guns firing. When the war began, the two guns were on the island of Borkum; after being used briefly in the West Wall, they were installed permanently at Calais.

The overhanging shield, known as the Todt front, had the shape of an inverted stairway and provided protection from bombs. It also deflected shells to the side.

Below: the Oldenburg Battery's concrete bunker, 15 meters high and over 80 meters wide, afforded the guns and crews ample protection from enemy fire, but what with their size, they were also easy to see from a long distance away.

A 28 cm SK L/50 gun of the Grosser Kurfürst Battery, raised and ready to fire. Note the two screws holding the mantle to the inner barrel. In the spring of 1944, work began on the construction of two more batteries, armed with 28 cm SK L/50 guns, near Knokke and Rozenburg. Whether the batteries had been completed at the time of the invasion has not been determined to date.

28 CM SHIP CANNON L/50

The 28 cm SK L/50 was a further development of the 28 cm SK L/45 of 1909, which served as the primary weapon of the Imperial battle cruisers "Moltke", "Göben" and "Seydlitz."

The gun was already modernized during World War I after the experience gained in the sea battle of the Dogger Bank in 1915.

The range of elevation was extended, and the barrel length was increased by five caliber lengths to L/50.

The gun performed considerably better as a result of being lengthened, and was lightened as well. The barrel still weighed 77,600 kilograms with a length of 14,150 mm. The exact caliber of the screwed-in inner barrel was 283 mm. The barrel was of the self-bearing type, meaning that it could absorb the stress of firing by itself.

The mantle was attached to the barrel in only two places and simply served to limit the overhang of the barrel to an allowable extent.

In addition, it increased the explosion safety, meaning the barrel's opposition to explosion when strong charges were ignited in the barrel.

The considerable play between the parts of the barrel allowed a simple changing of the inner barrel when it was shot out or damaged.

The gun had a wedge breech and was opened and closed mechanically with a crank. The recoil and recuperation system was divided between two hydraulic braking cylinders above and two pneumatic recuperators below the barrel. The air pressure in the recuperators, which also held the barrel at its elevation, had to be checked constantly, for which purpose manometers were attached to the barrel cradle.

The 28 cm SK L/50 was utilized by the Navy on the pocked battleships of the Deutschland class in double or triple turrets, and in single mounts in the "Grosser Kurfürst" shore battery.

In the coastal defenses, the barrel was mounted on the C/37 turning mount, and this in turn on an impervious supply bunker.

For protection against shot and bomb attacks, the guns were fitted with an armored turret of 15 cm steel plates.

Their traverse field of 360 degrees made the Grosser Kurfürst battery one of the most effective on the channel. Unfortunately, the barrels could no longer be replaced because of the war situation in 1944.

For this reason the range decreased from 30,200 meters for the L/4.4 explosive charge (weighing 284 kg) with a charge of 108 kg in 1940 to 26,000 meters in September of 1944.

The charge consisted of a main charge of 70 kilograms of RP C/12 powder contained in the cartridge case, and 38 kilograms of powder in the pre-cartridge of cellulose paper with an artificial silk cord.

Used as fuses in all German cartridges were C/12 (nA) percussion fuses.

In addition to the L/4.4 explosive shell, the L/3.6 bMdZ mHb percussion shell and the L/3.2 mBdZ mHb percussion shell could also be fired.

This picture shows the strength of the armored turret and the screws that hold it together. Under the gun barrel, the camouflaged Turret I of the next Todt battery can be seen. (BA)

Right: Camouflage nets are hung over the fourteen-meter gun barrel, the mouth of which is protected by a sack. (BA)

Two artillerymen examine the mouth of the 28 cm gun after a duel with British long-range guns. The barrel shows slight shrapnel damage. (BA)

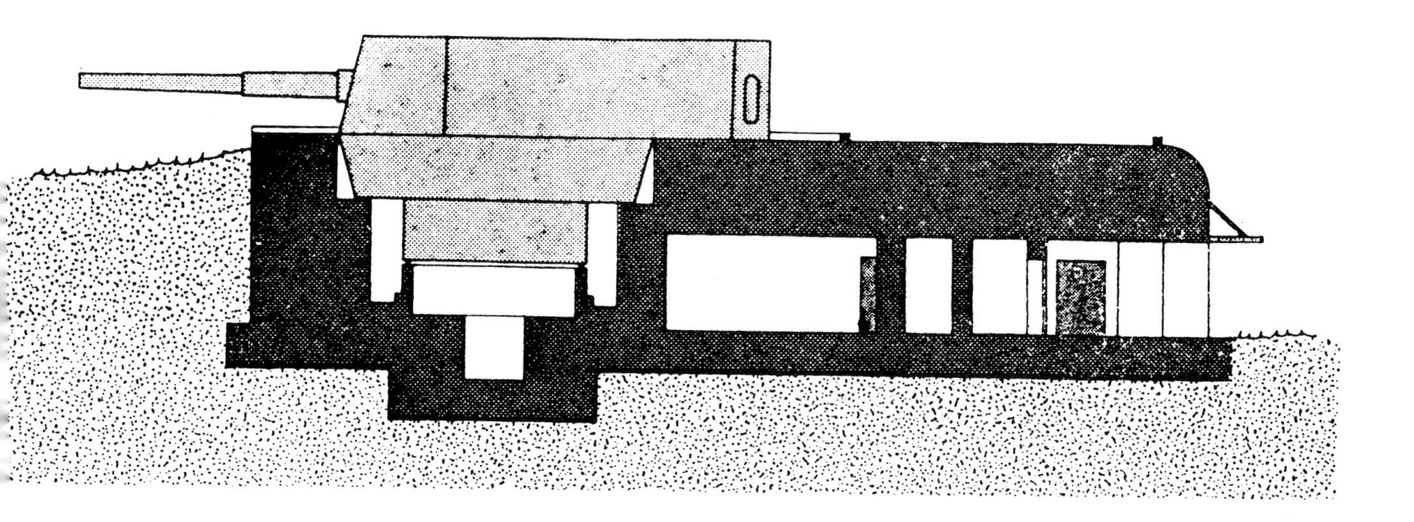

Gun Bunker, Grosser Kurfürst Battery

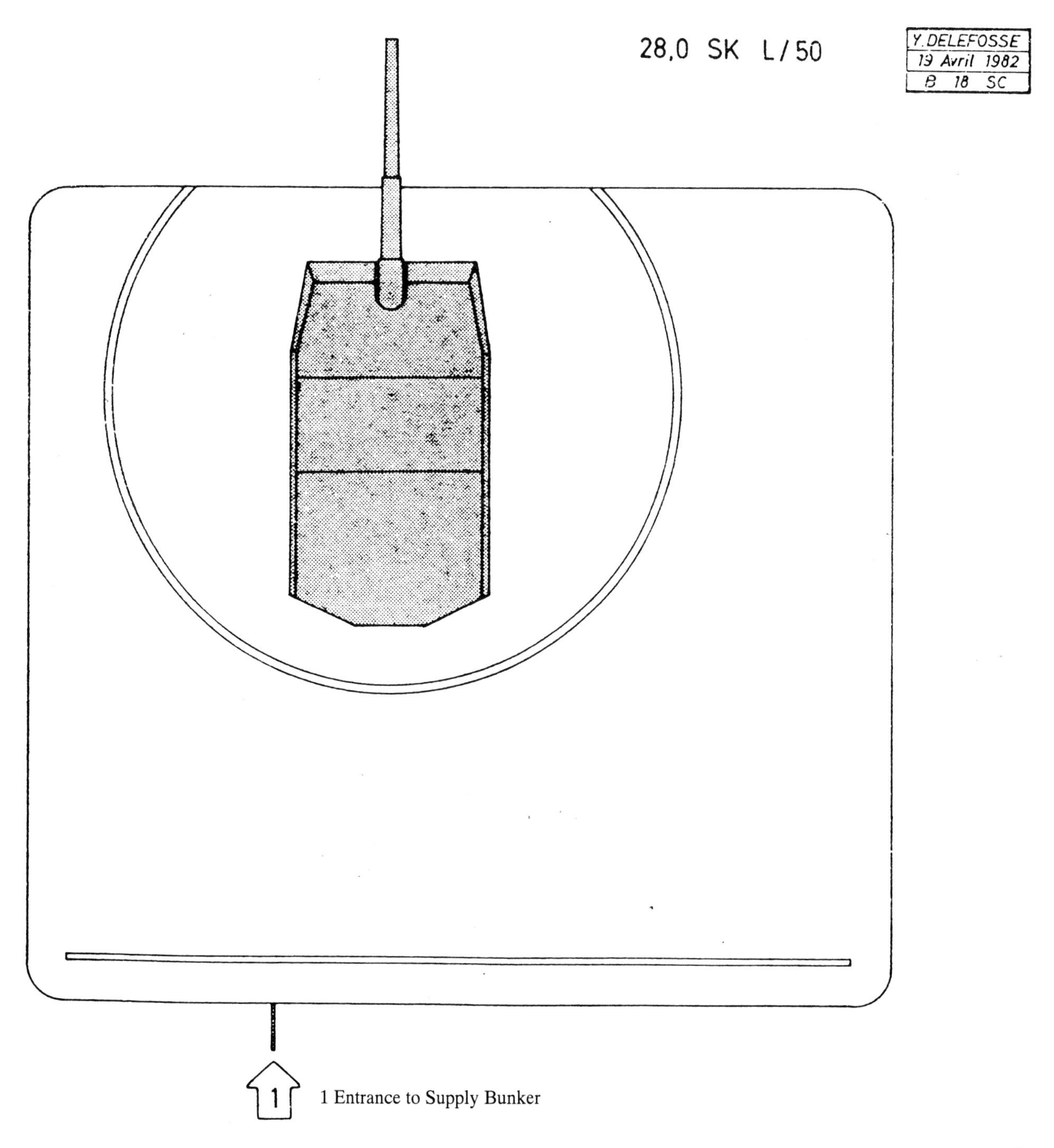

Left: The bedding ring of a 28 cm gun. This ring was installed at the lowest point of the turret shaft in the bunker, and the shell and cartridge elevators rested on it. Above is the gun-turret socket, about 1/3 of which extended out of the bunker top. The free-standing part was protected by a barbette or ring of armor.

Lower left: The shot-value receiver of a 28 cm SK L/50. By means of this, the fire-control calculator at the command post communicated the calculated data, such as the shot direction, distance, and angles of elevation and traverse. The K1 set the elevation gear and the K2 the traverse gear according to these values.

Below: The roof armor or barbette screwed onto a 28 cm SK L/50 gun.

Of the several hundred bombs dropped on the Grosser Kurfürst Battery, only eight hit the gun turrets. None of them knocked a gun out of action, though the roof armor of Turret I was raised slightly. The entrance to the bunker is still barricaded with boxes and ammunition containers filled with sand. (IWM)

Left: The Grosser Kurfürst Battery area after heavy bomb attacks. In the right foreground is Turret I, and Turret II can be seen in the center of the picture. (IWM)

In front of Turret IV, a Canadian soldier examines what remains of a bomb after the battle. The gun turret is still draped with camouflage nets. (IWM)

Gun I of the Friedrich August Battery on the Channel coast near Wimille in September 1940. The gun is still in an open firing position and barely camouflaged with nets.

30.5 CM SHIP CANNON L/50

The 30.5 cm SK L/50 was developed by Krupp in 1906 for battleships (beginning with the Helgoland) and battle cruisers (beginning with the Derfflinger). The gun was very solidly built and, like the 28 cm SK L/50, had a wedge breech with a crank and self-cocking mechanical firing.

The barrel, more than 15 meters long, had a progressive rifling, with one rifling making a full circuit in a distance of 40 to 50 caliber lengths.

The inner barrel consisted of replaceable parts and was likewise screwed in place.

The initial velocity for the L/3.6 explosive shell (which weighed 250 kg) and a powder charge of 149 kg was 987 meters per second.

With the L/3.8 explosive shell (weighing 405 kg) and a powder charge of 127 kg, the velocity was 768 meters per second.

In the Reichsmarine, these barrels, weighing 68,000 kilograms, were installed in bedding devices and stationed on the island of Wangerooge.

In 1940 the Navy divided the battery into two, the Friedrich August and the Schröder, each with three guns.

After a short stay on the West Wall, the Friedrich August battery was stationed near Wimille in September of 1940.

With a maximum shot range of 32,500 meters, the guns could reach the harbors of Folkstone and Dover.

The battery served for the most part as an instructional battery for the training of naval artillerymen.

The elevating and shell-inserting apparatus could be operated electrically. For this purpose the battery was equipped with a 220/380 V 40 KVA generator aggregate.

Traverse aiming had to be done manually. During training, all gun movements were made by hand.

The barrel of the 30.5 cm SK L/50 gun, 15 meters long, has two ring mantles screwed on. The lighthouse of Wimereux can be seen in the background. (BA)

The photos on this page were taken during a firing drill. Under the gun are transport carts with shells, pre-cartridges and main cartridges. Used cartridges lie in the foreground, cooling off. (BA)

A new transport cart is wheeled under the gun's elevator. Next to the shell is the main cartridge with 89.9 kilograms of powder; below is the pre-cartridge with 67.1 kg. The two form a complete charge of 149 kg of powder. (BA)

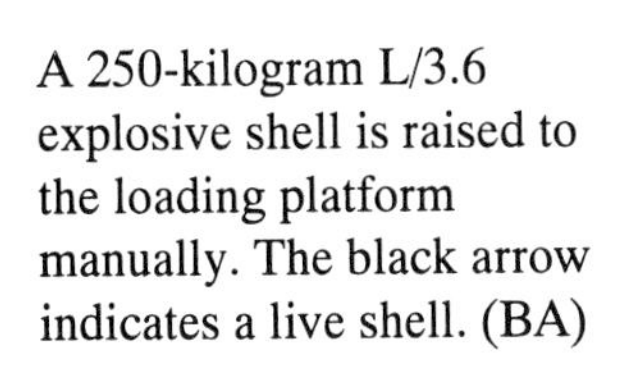

A 250-kilogram L/3.6 explosive shell is raised to the loading platform manually. The black arrow indicates a live shell. (BA)

Left: On the loading platform, shells and cartridges are moved ahead of the breech and pushed into the gun by hand. Electric shell injectors and elevators were installed only when the gun bunker was built. (BA)

Below: An impressive picture of a 30.5 cm shell being fired.

After firing, the barrel is immediately returned to the loading position of +2 degrees. When the breech is opened, the cartridge is automatically ejected and then removed by the loader.

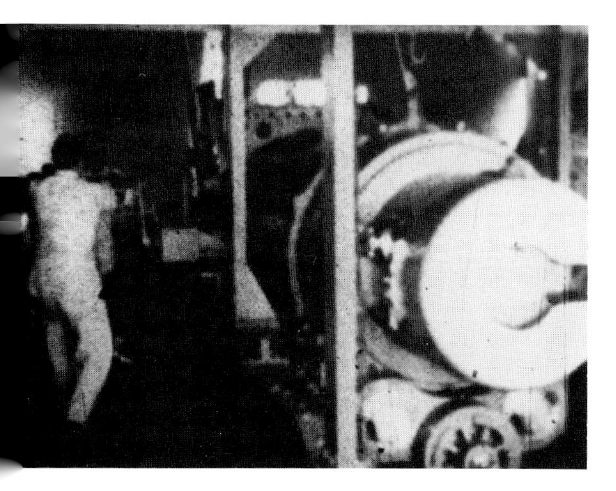

Above: Two artillerymen are cranking the gun back into its loading position.

Right: A 40-kilowatt generator that supplied the elevation gear and the shell elevators of the guns.

After a shot is fired, the barrel is cleaned by all hands, under the direction of the gun leader, Ofw. Zeidler.

The bunker wall of Turret II has been hit by numerous armor-piercing shells. But every turret had to be captured by Canadian troops in close combat.

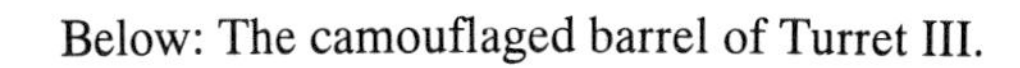

Below: The camouflaged barrel of Turret III.

Gun Bunker, Friedrich August Battery

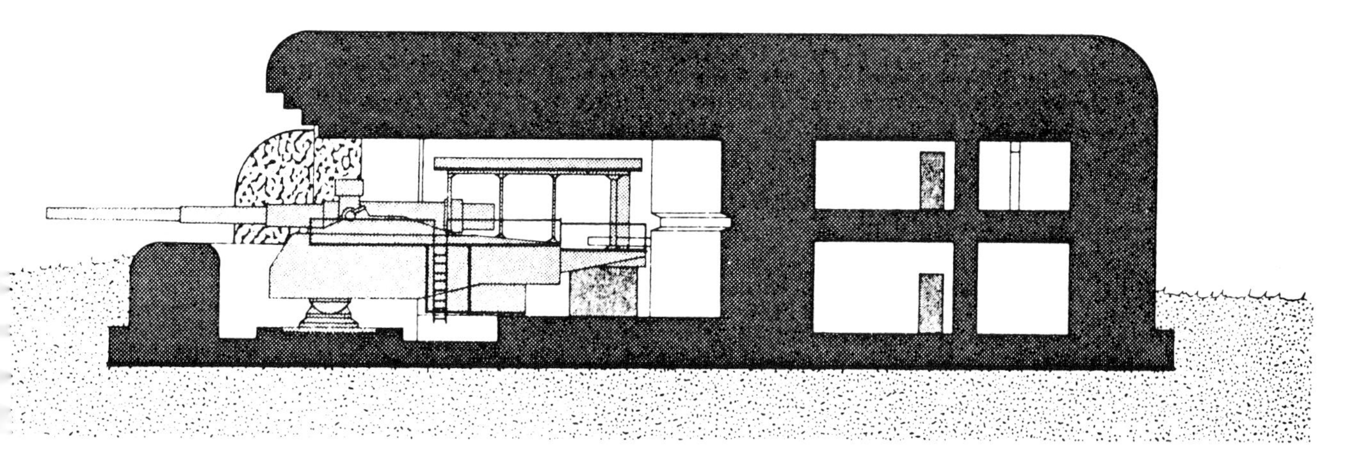

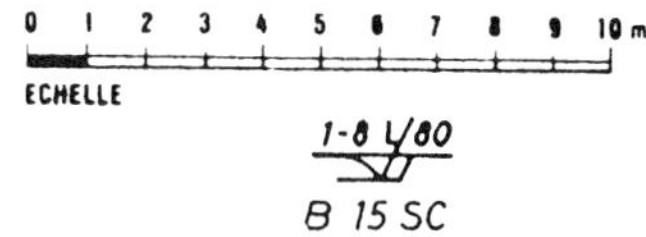

Left: Turret III fires on a convoy in the Channel.

Right: Turrets II and III of the Friedrich August Battery after being stormed by Canadian troops. Shortly afterward, all the turrets were blown up. (IWM)

38 CM SHIP CANNON C/34 (L/51.7)

The 38 cm SK C/34 guns were mounted in double turrets on the battleships Bismarck and Tirpitz. The replacement barrels intended for the ships were mounted in C/39 bedding devices in Norway, Denmark and France for use in coastal defense.

The 38 cm gun had a screwed-in inner barrel with a mantle, and two tightened ring mantles strengthened the barrel parts which were exposed to the greatest stress.

The barrel was 19.63 meters long, had 90 riflings, and weighed 105,000 kilograms including the breech.

The breech, as before, was a self-cocking wedge type with mechanical firing; it was opened and closed by a hand crank.

The breech wedge formed the resting place for the base of the cartridge. This wedge ran in the grooves of the wedge carrier and could be opened either to the left or the right – depending on which side of the ship's turret, for which the guns were originally intended, the gun was to be installed in.

A closing plate limited the closing movement. Firing was possible only with a completely closed breech and was done by a rod with a trigger line.

The shells, cartridges and pre-cartridges were transported from the ammunition areas to the loading stage by elevators.

There they were placed on loading carts and moved directly behind the breech. Now they were inserted into the barrel by the electric loader, while the barrel was at an elevation of four degrees, and then the breech was cranked shut.

After that, the barrel was moved into its firing position and the shot was fired by means of the trigger line.

Ignition was achieved by the percussion fuse C/12 nA. which was screwed into the cartridge. This fuse had a diameter of 33 mm for the 38 cm cartridge.

On firing, the striker in the breech of the gun pushed the base of the igniter in without penetrating it and thus ignited the percussion cap.

This flash of fire now entered the ignition bell, the actual powder chamber of the cartridge, and ignited the powder there.

Three types of shells were fired:

– The 38 cm L/4.6 explosive shell with ballistic sheath and nose fuse; this shell, 175 cm long, was the standard shell for the 38 cm guns.

– The 38 cm L/4.4 mHb explosive shell with base fuse, 168 cm long.

– The L/4.4 armor-piercing shell, likewise with sheath and base fuse. This shell was also used in action against armored ships.

All these shells weighed 800 kilograms and were fired with a brass cartridge (110 kg of RP C/38 powder) and a pre-cartridge of 101 kg powder. The shells, with an initial velocity of 820 meters per second, had a maximum range of 42,000 meters.

The upper picture shows a 38 cm SK C/34 in Turret I of the Todt Battery.

Left: The gun socket and turning ring of the 38 cm gun. This socket served as a mount and turning axis for the gun turret. In addition, the rear part of the gun turret rested on two rails that encircled the gun chamber or platform.

This picture shows the barrel of the 38 cm gun, weighing over 100 tons, being lifted onto the firing platform. in the foreground the frame of the armored turret stands ready.

Below: The gun and socket are mounted, and now the turret is coated with dark yellow camouflage paint. Under the turret is the gun socket, which is screwed onto a flattened cement cone. On the left side, half as high as the turret, the running rails on which the rear section of the turret rested can be seen.

Gun Bunker, Todt Battery

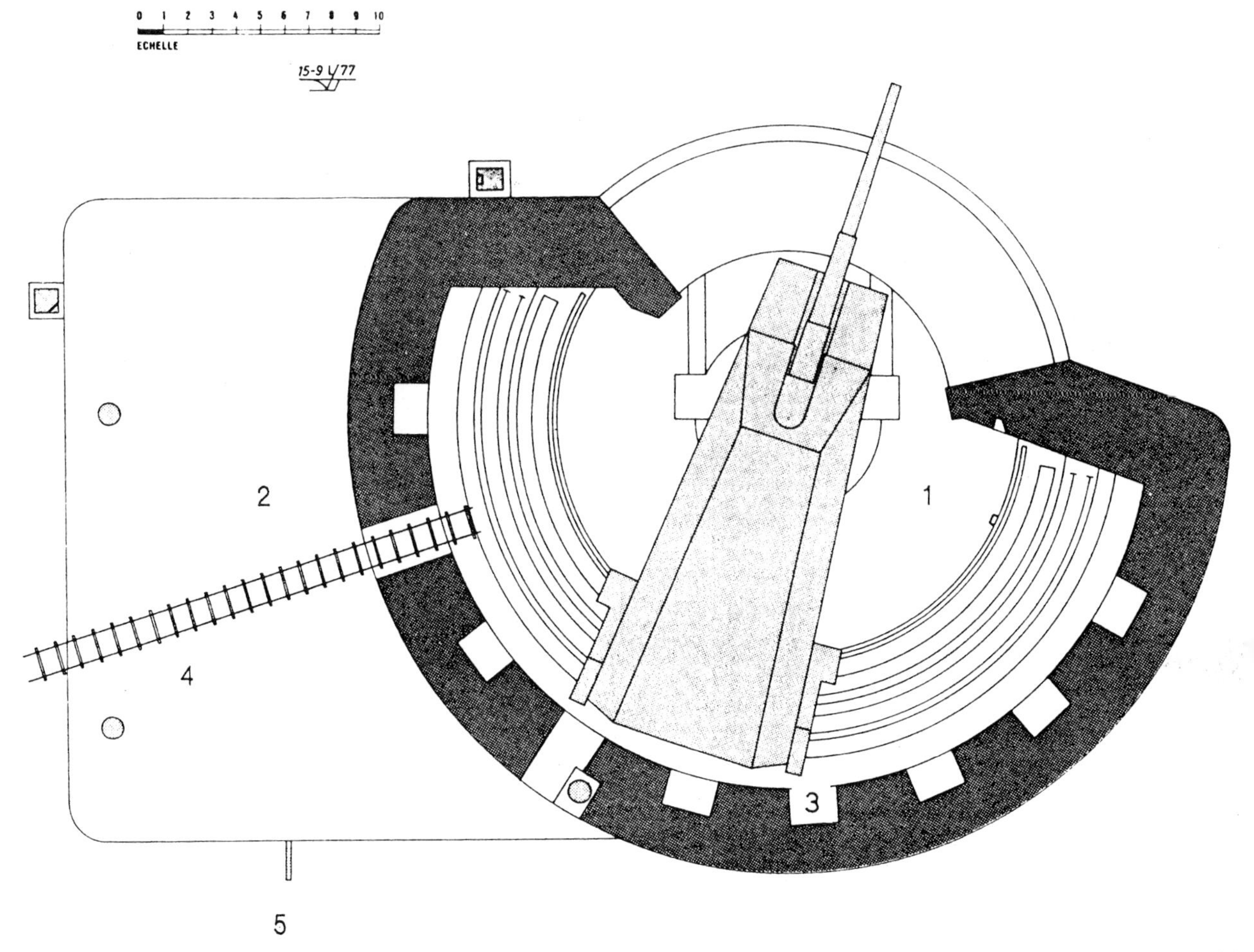

A drawing of the gun platform. 1. platform, 2. supply bunker, 3. ventilator, 4. rails of the narrow-gauge railway on which th ammunition could be brought directly into the turret, 5. entrance to the supply bunker.

Below: At the back of Turret I, a house was painted for camouflage. The ventilators were cleverly integrated into the pictur (BA)

The breech of the 38 cm SK C/34. The recuperator can be seen above it. The duct that piped powder smoke directly outside is attached to the right side of the breech.

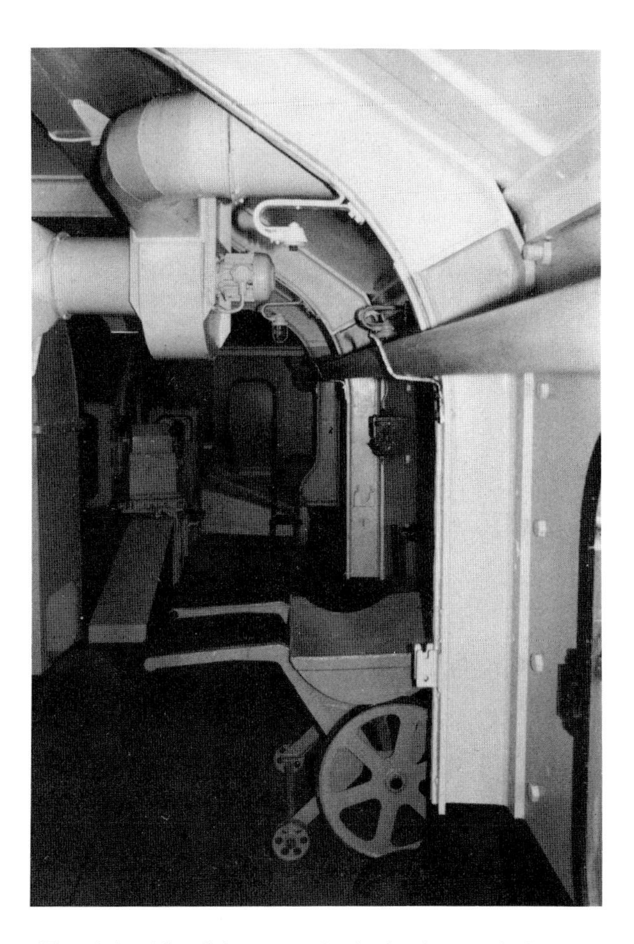

The right side of the turret. In the background, the gun cradle of the firing platform can be seen. In the foreground is a cartridge cart, on which hot cartridges were taken away to cool.

Below: The elevation gunner's position with handwheel and elevation gauge.

Below: The cartridge cart in front of the chute by which used cartridges were removed from the turret.

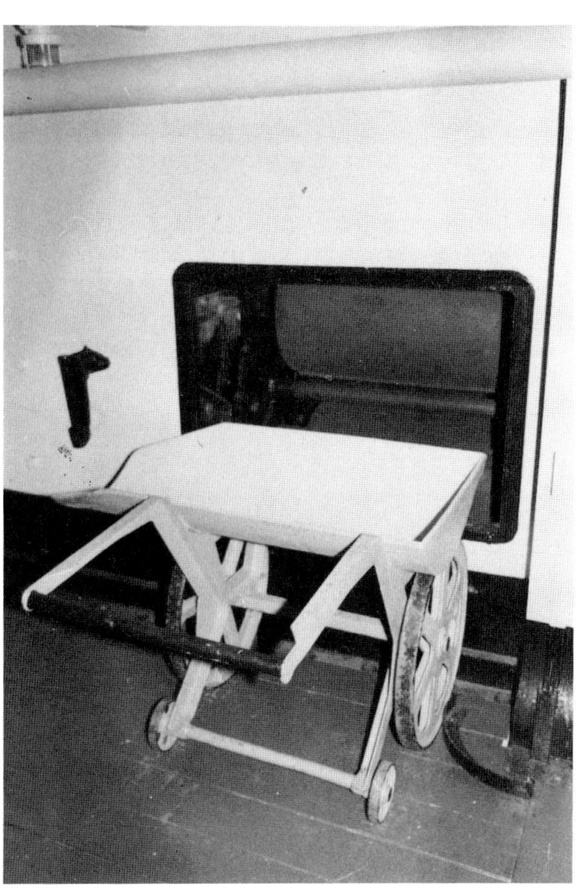

The back of Turret III with its built-on supply bunker.

Below: A cutaway drawing of the supply bunker. The ammunition rooms were on the upper level, the machine rooms on the lower.

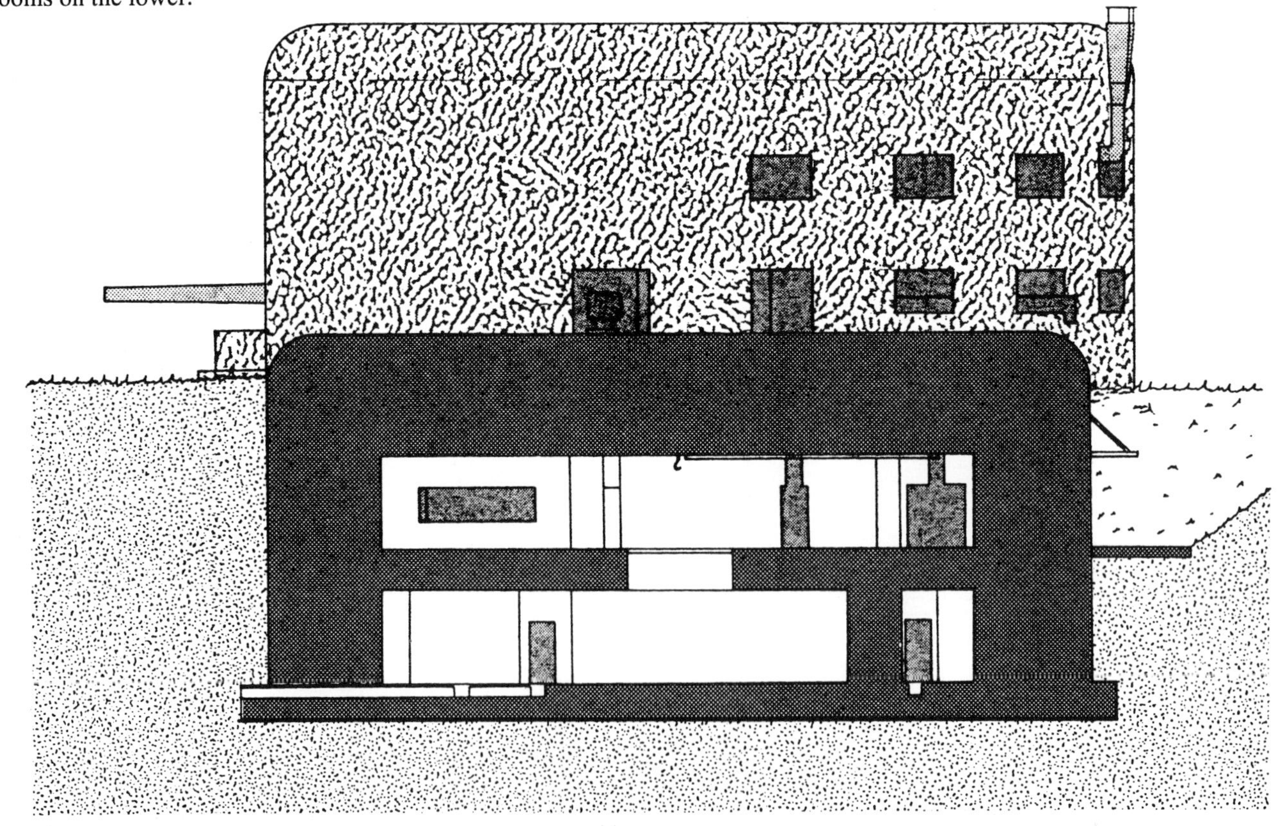

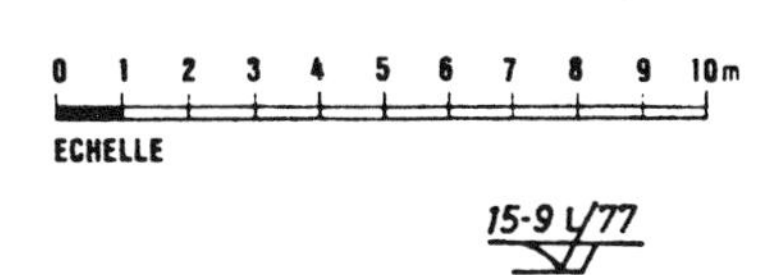

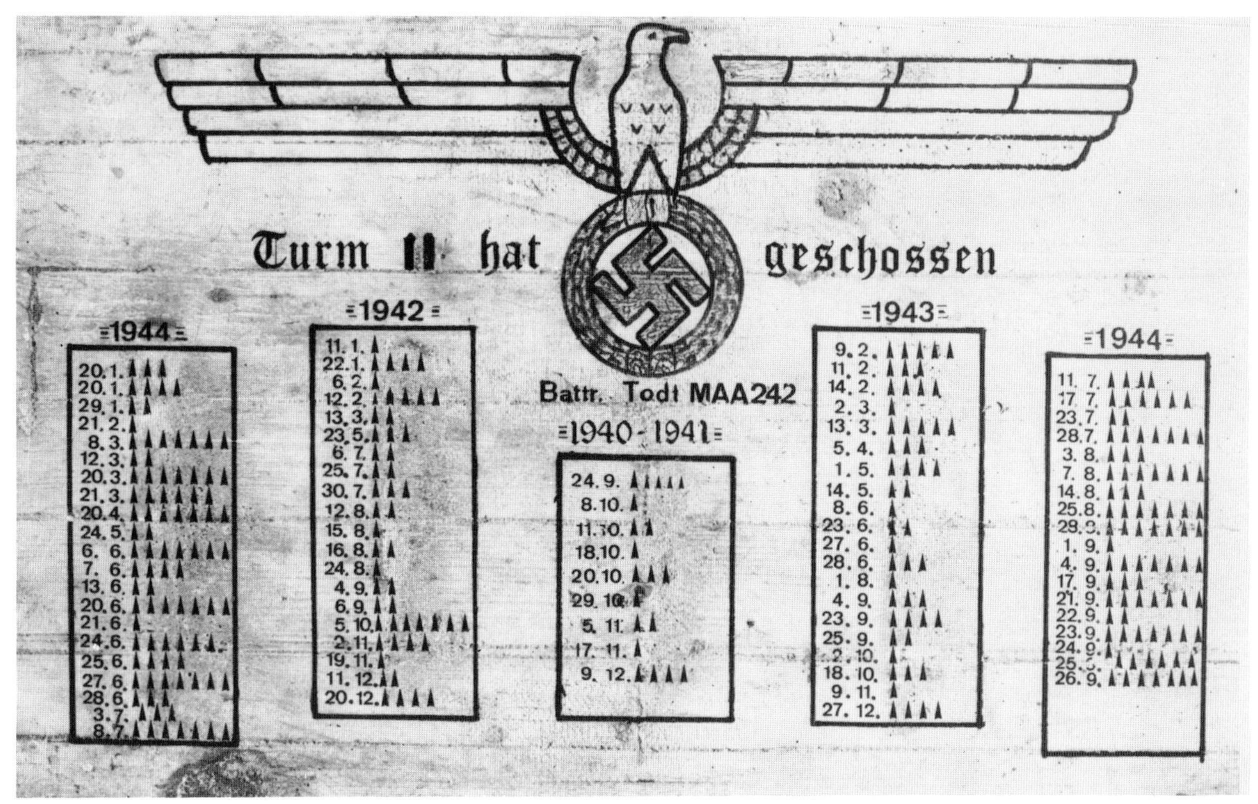

The firing tables for Turret II. Every gun had these tables painted on the bunker wall in the entry area.

Right: A 30.5 cm L/3.8 explosive shell with base fuse and ballistic sheath: Weight 405 kg, length 1.23 meters. Beside it is a 38 cm L/4.4 explosive shell with base fuse and ballistic sheath: Weight 800 kg, length 1.68 meters.

Below: This photo shows Turret III while firing.

Left: The interior of a gun turret after the Canadians captured it. Part of the smoke duct and the breech have been removed.

Below: Artillerymen of Turret II show the size of the 38 cm gun barrel in this photo.

Below: Turret III in August of 1945, after being destroyed. All the turrets and the command post had been prepared for being blown up, but only the command post was destroyed shortly before the Canadians captured it. In August of 1945, two Frenchmen accidentally set off the explosives and lost their lives as a result. German prisoners removing mines in the vicinity saw the steel covering of the gun position, weighing several tons, fly several hundred meters through the air.

This postwar photo shows the steel plates welded onto the sides of the turret to cover the open space when the turret was turned to the side. In spite of them, flamethrowing tanks were able to set fire to the interior of the turret.

Below: The last photos of the Todt Battery's 38 cm gun, taken by members of the German mine-removal command of Griz Nez.

40.6 CM SHIP CANNON C/34 (L/52)

In 1934 the Krupp firm, under contract from the Navy, began to develop and build the 40.6 cm ship cannon.

This gun was intended to be the primary armament for the 56,000-ton battleships of the "H" and "J" types.

The caliber was the largest allowed by the Washington and London naval agreements.

On account of the war and manufacturing situations, though, the planned ships were never built.

The eleven guns that had been ordered were mounted in C/39 bedded firing devices and divided among two batteries in Norway and the Schleswig-Holstein Battery near Sangatte.

The 40.6 cm ship cannon was structurally identical to the 38 cm SK C/34, differing only in the caliber.

The barrel, 21,000 mm long, likewise had ninety riflings. These guns fired the L/4.8 explosive shell with nose fuse and sheath, the L/4.6 explosive shell with base fuse and sheath, and the L/4.2 armor-piercing shell. All three shells weighed 1030 kg and, with an initial velocity of 810 meters per second, attained a maximum range of 42,800 meters.

Above: The 40.6 cm C/34 gun of Caesar Turret at its maximum elevation of 52 degrees. The bunker is still under construction and partly scaffolded with wood. A group of visitors is just leaving the bunker.

Below: This drawing of a gun bunker shows its size. Behind the gun chamber were three stories with barracks, ammunition magazines and machines. At the left is the entrance for trucks and the narrow-gauge railway.

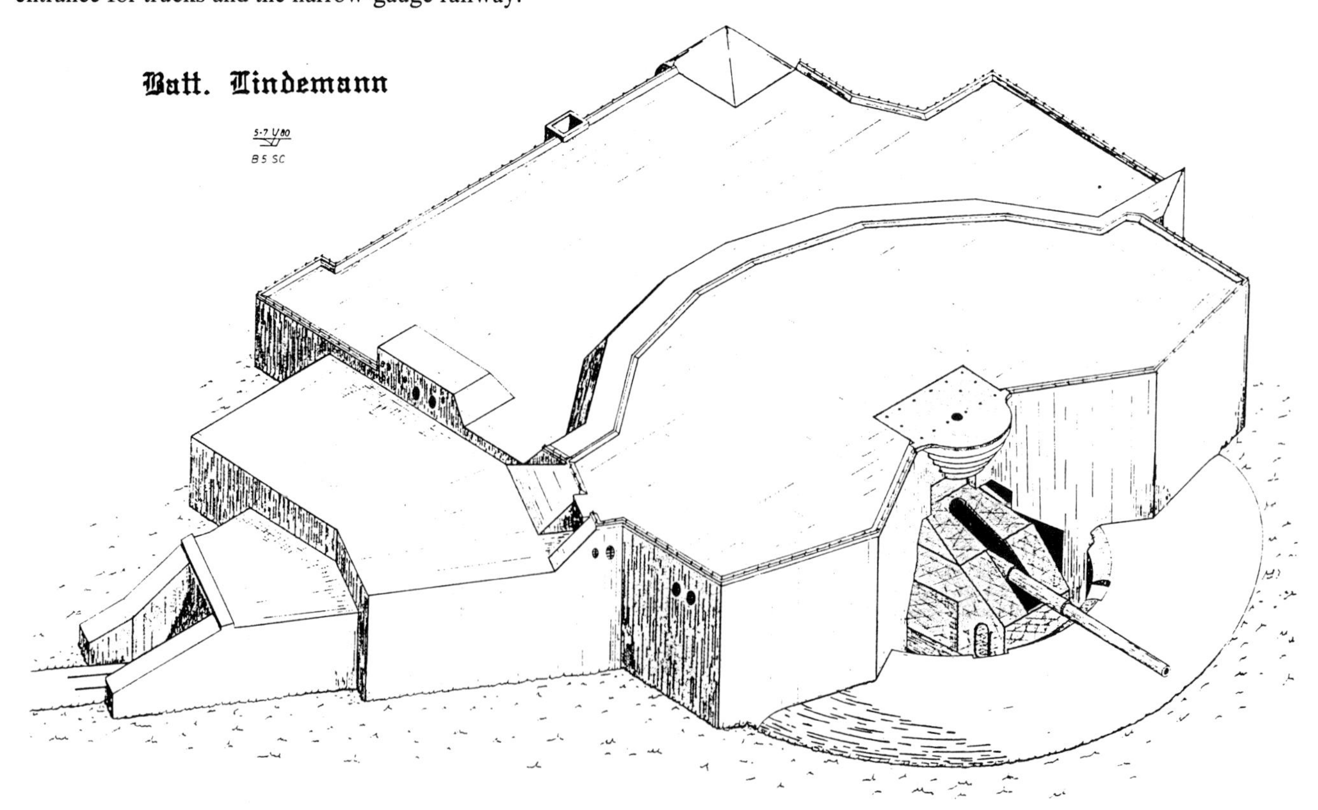

Above: This photo shows the great size of the 40.6 cm C/34 gun. In the center is the shield with which the barrel is anchored in the gun cradle of the C/39 firing platform. (BA)

Right: The barrel is lifted onto the firing platform by two cranes. (BA)

Above: This picture shows the bunker of Caesar Turret nearing completion. A primary carrier of the concrete cover is just being moved into position. The site of Bruno Turret can be seen in the background. (BA)

Below: The scaffolding of the shield has been removed.

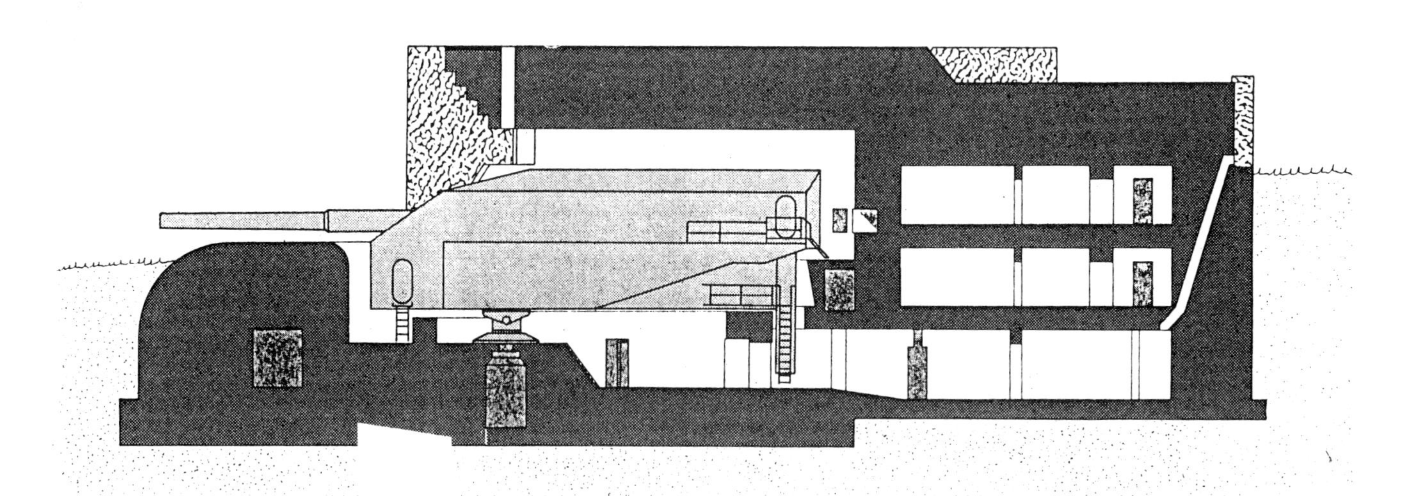

Above: A cutaway drawing of a gun bunker. The concrete walls and roof, four meters thick, were not penetrated despite being hit by many bombs. Only the Bruno Turret was damaged on September 3, 1944, shortly before the battery was captured. A shell from a British railroad gun hit its elevating gear.

Right: The gun crew climbs into the turret. At right, next to the steel door, is the cartridge case ejector, through which used cases were moved out of the turret.

The narrow-gauge railroad line, seen at the entry to the bunker. By using the hoist, the bar of which is seen on the ceiling, ammunition, spare parts and fuel could be moved into any room on conveyors.

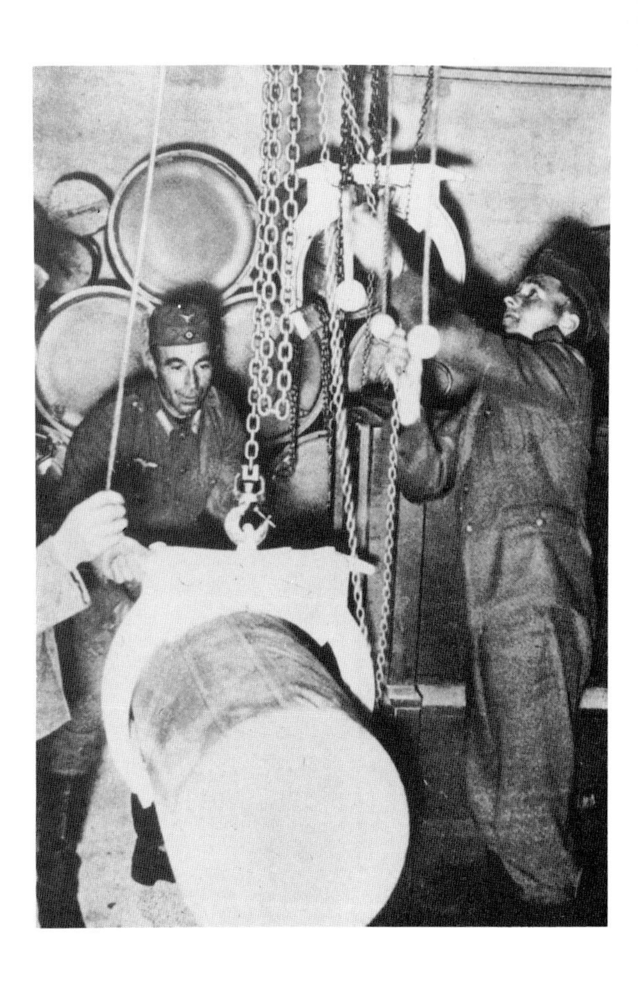

A pre-cartridge weighing 101 kilograms is stored in the powder magazine.

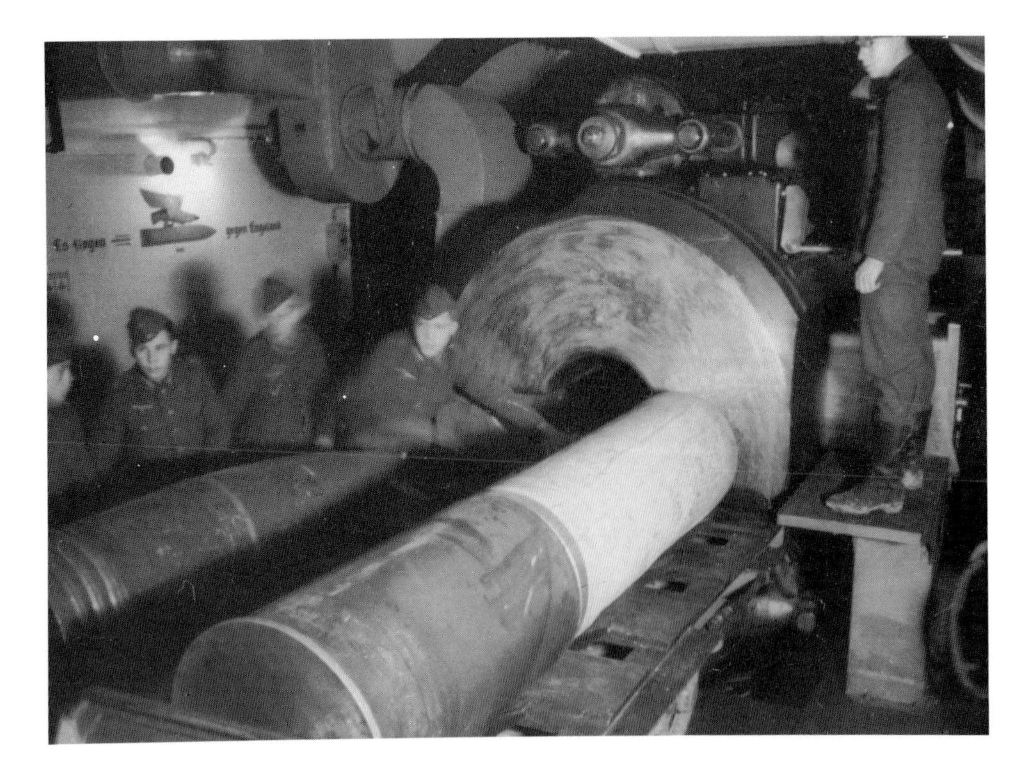

Left: Loading the gun in Caesar Turret. The shell, pre-cartridge and main cartridge lie on the loading cart. The loaders are rolling the shell toward the loading hole; it will then be pushed into the barrel by the electric apparatus. Behind the artillerymen the shot tables can be seen. On the right side stands the K1 on a platform, ready to close the breech. (BA)

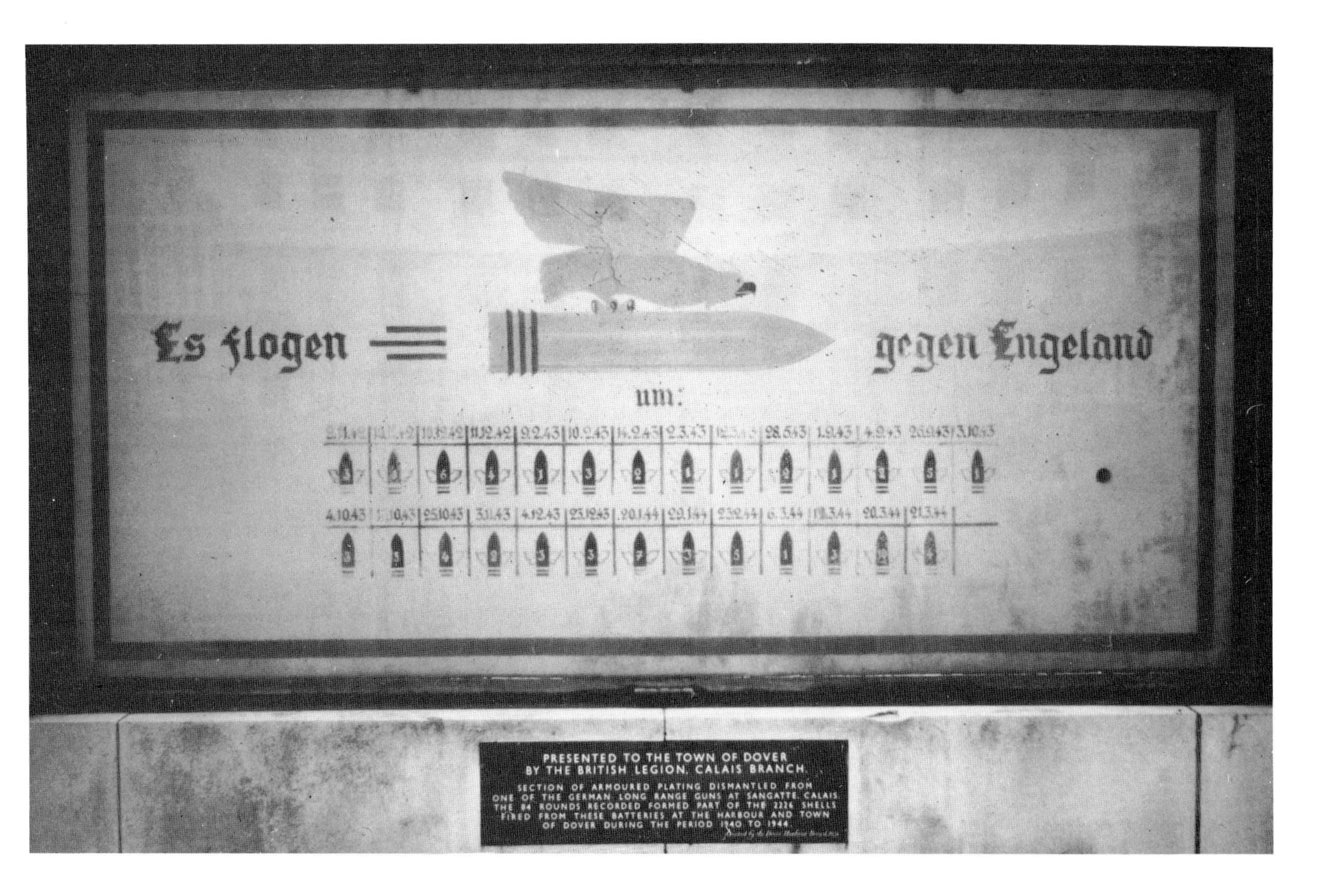

Above: the shot tables of Caesar Turret, which stand on the Promenade at Dover today.

Lower left: The naval artillerymen's two uniforms, at left the field gray version, at right the white type, which included asbestos fibers to make it less combustible. The two uniforms now staid in the memorial hall of the Lindemann Battery, in the Calais City Museum.

Below: A platoon of the 800-man crew marches past the bunker, hung with camouflage nets, for training.

FIRE-CONTROL DEVICES

OPTICAL RANGE FINDERS

The long-range guns could fire on targets at ranges up to more than 40,000 meters. For targeting, the batteries were equipped with range-finding or targeting devices made by the Zeiss firm. They ranked among the most precise and best-performing devices of their time.

These measuring devices worked according to the mixed-image principle. The target observer saw two identical pictures of the intended target in the ocular. When the two pictures were brought together by turning the handwheel, he could read the distance on a scale.

The farther apart the two measuring beams of the device were, the more precise the results of the measuring were. The primary command posts of the heavy batteries were equipped with ten-meter range finders. The secondary command posts were usually equipped with seven- or four-meter devices.

The captured French batteries, Créche I and Bastion II, retained their range finders of the Indicateur de Gisement type (1929 model) at their primary command posts, while each secondary post had a two-meter device.

The aiming devices had 400-degree graduation as did the guns, plus double optics and very good yellow filters.

On account of light and friction losses as well as tolerance limitations, these aiming devices naturally had their technical limits. In addition, weather and light conditions had a great effect on their possible use.

The measured values were transmitted electrically to the fire-control calculator (long-base parallax calculator).

After the target observed had adjusted the range finder manually to attain a "good measurement" at a point or sustained contact, the values were transmitted to the calculator either continually or individually.

The fire-control calculator processed the values on mechanical gear drives. There were drives for types of ground calculation, angle-function drives, differentiation and integration drives to determine minimum and maximum values, as well as coordinate adapters.

Constant values, such as the distance of the gun from the range finder, could be stored in the calculator permanently. The calculated aiming values for the guns were transmitted to them either by telephone or by loudspeaker.

Firing was also done according to long-base measurement, called "measurement" for short. In this process, the firing battery and the target were each sighted through two theodolites, instruments for horizontal and vertical angle measurement.

The location of the impact, in relation to the target, was read from the gradations in the objective of the range finder. Evaluation required considerable calculation, to be sure, but provided very exact results.

The massive turning dome of the Todt Battery's electric range finder at Cape Gris Nez. In cooperation with the subsidiary fire-control posts, it could cover an arc of 342 degrees. (BA)

Right and below: Two pictures of the Oldenburg Battery's six-meter range finder. The target observer or targeting officer is aiming at the target. With his right hand he brings the images of the target toward each other to determine the exact distance.

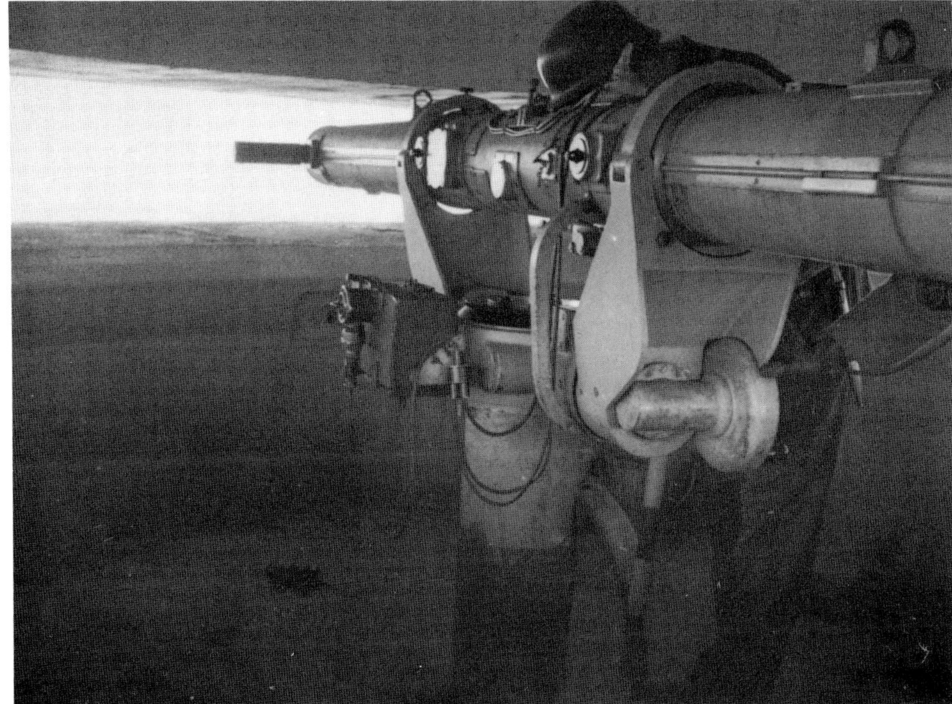

In the photo at left, the cables at the front of the device that link it with the fire-control calculator can be seen. The steel helmet is surely used here only for the photograph; in reality it was only a disturbance.

Below: The ocular of a 2.2 meter device. To the left is the handwheel that brings the two images toward each other; to the right of the ocular is the "good measurement" handle, with the two levels of point or steady contact. Under the bar is a scale from which the distance could be read.

Left: Naval artillerymen move part of the fire-control calculator into the control post of the Lindemann Battery. In the background is the steel cupola of the ten-meter range finder. (BA)

Below: The optics of the ten-meter device. In case of damage, this part of the optics could be changed easily by hand.

Right: The M I Battery's fire-control post, in which a four-meter range finder is housed. The shape of the tower resembles that of a ship's bridge.

Below: A cutaway drawing of the Lindemann Battery's fire-control post. 1. ten-meter range finder, 2. Würzburg Giant FuSe 65 device, 3. calculator area, 4. machine and fuel storage area.

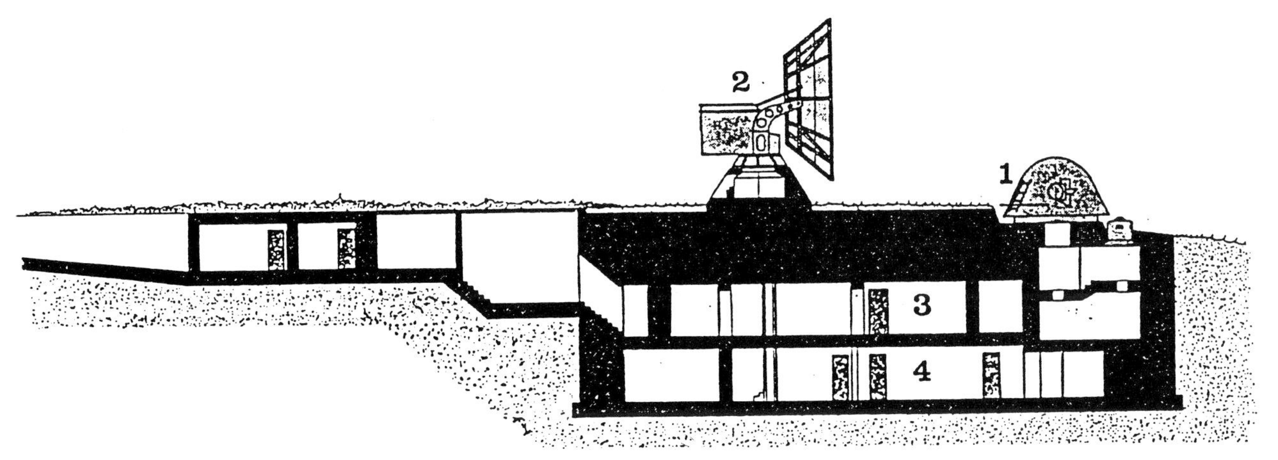

Right: The upper part of the steel cupola is placed over the ten-meter range finder by a mobile crane. (BA)

A 2.2-meter range finder in the fire-control post. The windows of the bunker could be folded down if necessary to afford a better view.

Below: The main fire-control post of the Creche I Battery. In the upper story is the Indicateur de Gisement range finder. The protective cover of a periscope through which the battlefield was observed can be seen below it.

RADAR ORIENTATION

In addition to using optical measurement, the Navy began in 1934 to use electromagnetic waves, now known as radar, for targeting.

The development of electronic radar devices (FmG) by the GEMA, Telefunken, Lorenz and Siemens firms made it possible to measure a target's distance, elevation and traverse angle. These values could be transmitted to the guns via the fire-control calculator as a basis for firing.

The first radar devices (FmG 39 G), referred to for purposes of concealment as De-Te devices, Decimeter-Telegraphy, Funk-E measuring devices or Em-II devices, reached the Channel in the summer of 1940.

The FmG 39 (gB) "Calais" device had an antenna measuring two by six meters. The upper half of the antenna was for receiving, the lower half for transmitting. The range for sea targets was some 15 to 20 kilometers, with a distance-measuring precision of 70 meters and a lateral precision of about +/- 3 degrees.

As of 1942, this device was known to the Navy as FuMo 1.

The last heavy battery, the Oldenburg Battery, received its FmG 39 G radar device in December of 1940.

The long-range batteries received the FmG 41 G (gA) or FuMo 51 "Mammut-Gustav", in a radar station near Onglevert, as a long-range searching device. The antenna, measuring ten by twenty meters, could be turned some +/- 50 degrees on a central mast, and its range was approximately 50 kilometers.

The naval artillerymen learned, by exchanging experiences with the Luftwaffe, to acquaint themselves with the Funksender 65 (FuSe) "Würzburg Giant" device at Blanc Nez.

As a result, the Lindemann, Todt, Friedrich August and Grosser Kurfürst batteries were equipped as of 1943 with a "Würzburg Giant" devise, designated FuMo 214 by the Navy.

The "Würzburg Giant" had a parabolic reflector 7.5 meters in diameter. By using a defocused, rotating dipole, this device attained a lateral targeting accuracy of +/- 0.2 degrees and an elevation accuracy of +/- 0.1 degrees.

Above: The "Würzburg Giant" 65 radar dish near Blanc Nez. The device stands on a concrete socket.

Below: A drawing of the 39 G (gB) radar device in its shrapnel box.

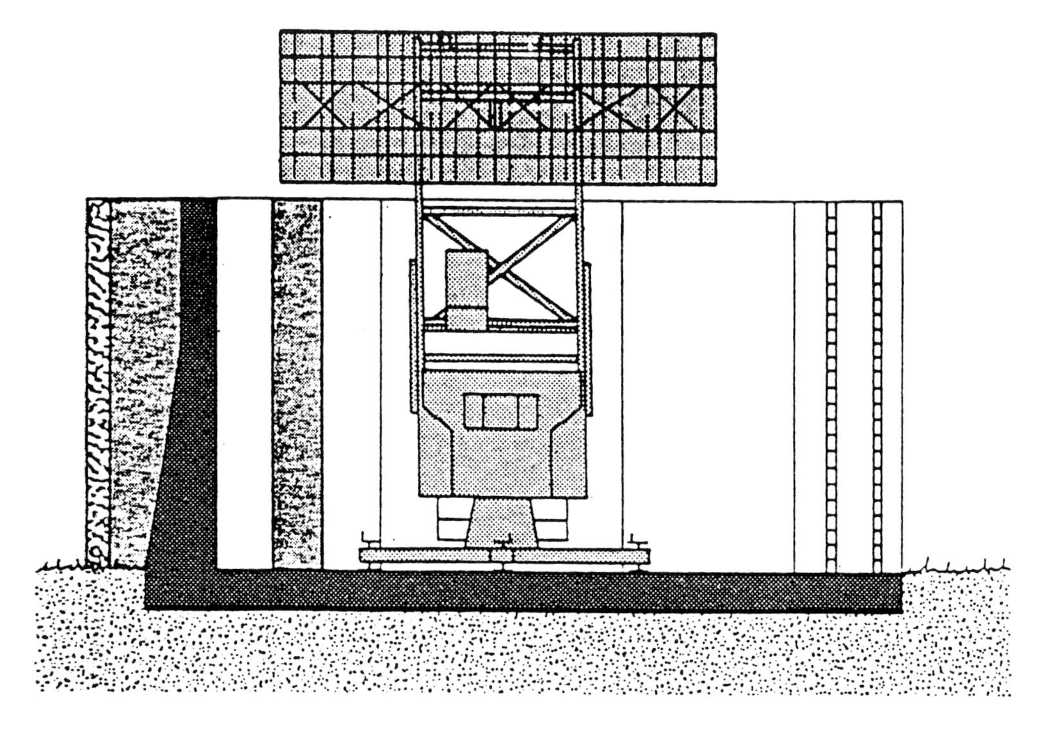

Left: A gun bunker of the 5th Battery, MAA 242, near Waringzelle at Cape Griz Nez. In the background is the concrete socket of the Type V143 bunker, on which the FuMo 21 was to be mounted.

In the summer of 1944, construction of a 15 cm battery with its own radar facility began near Gris Nez. The three 15 cm SK L/45 guns were to be linked directly with the FuMo 51 "Mammut" device. As Canadian troops advanced in August of 1944, the work had to be halted. Today one can still see the state of the work at this very isolated location.

Also worthy of note are the Luftwaffe and Navy flak batteries in the Boulogne-Calais area. All of them were equipped with FuMo 212 "Würzburg C" radar devices, and later with FuMo 213 "Würzburg D" types. These considerably smaller devices, though, could not be used for fire control of heavy artillery.

In all, the Navy's radar equipment was not utilized to its greatest possible extent. Only in 1943 was radar divided between intelligence and targeting functions, and schools for the new devices were founded even later.

A 41 G (gA) radar device on a shrapnel box near Onglevert.

A drawing of the 41 G (fB) radar device, with an improved pivot. These devices quickly replaced the earlier FmG 41 (gA).

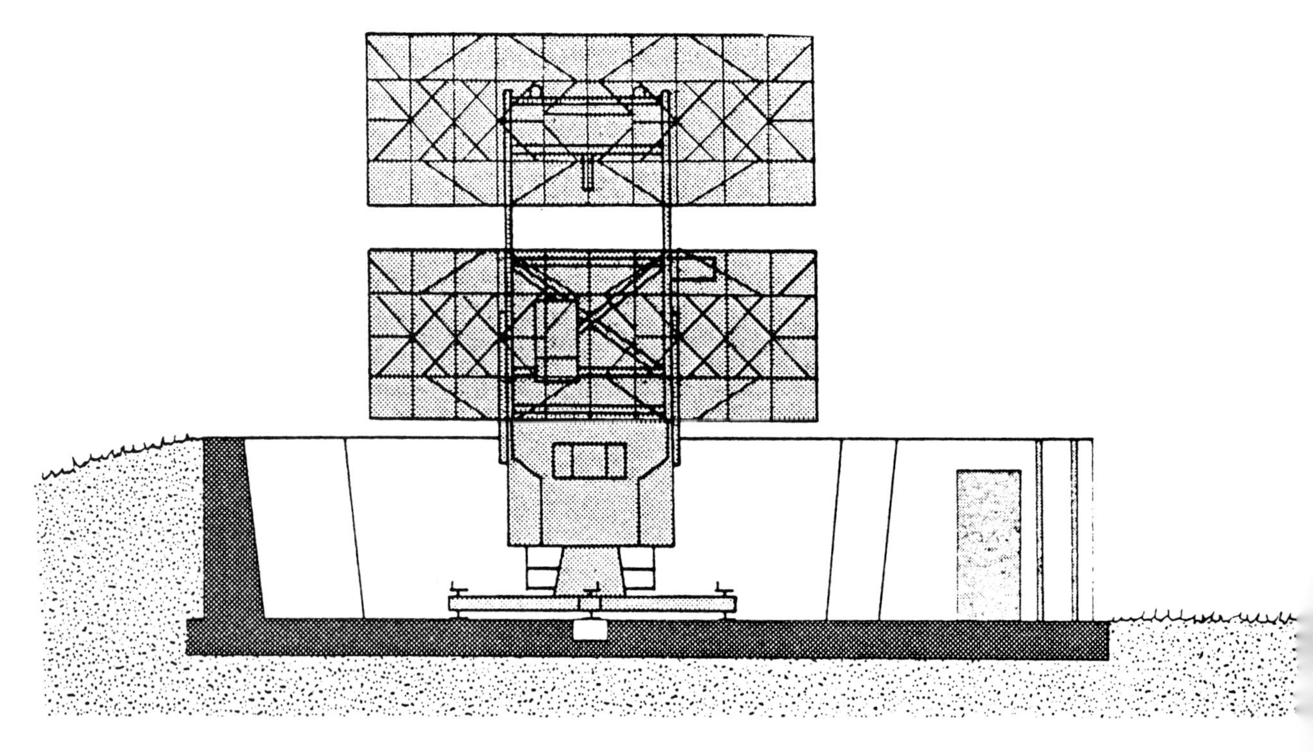

This FuMo 213 "Würzburg D" device was abandoned by a naval anti-aircraft unit.

HEAT-SENSING DEVICES

In addition to radar, the batteries had only spotlights and tracer shells to use for targeting during night combat up to 1941. In the spring of 1941, every naval artillery unit on the Channel was issued several heat-sensing or infrared devices.

These devices were based on the fact that certain elements change their electric resistance when their temperature changes. In the socket of a spotlight reflector, a bolometer (selenium cell) was installed in place of a source of light. By radiation from a source of heat, the resistance was at its smallest or greatest extent when the maximum of radiation was attained, or the bolometer was directed exactly at its target.

The result was shown in a Braun Tube, divided into traverse and elevation. By turning the handwheels, the points of the curves had to be moved into the centers of the tubes, and thus the adjusting controls indicated the target. In April of 1941, the MAA had 240 heat-sensing devices in positions near Zeebrugge, Ostende, Dunkerque, Blanc Nez and Cap d'Alprech, plus a subsidiary depot in Wimereux.

The Grosser Kurfürst and Oldenburg batteries are said to have sunk enemy ships by using these heat-sensing devices. Unfortunately, little is known about this very interesting aspect of fire control.

A 60 cm spotlight that was used as a heat-sensing device. The lens is turned straight up and the optics have a protective cover. In front of the operator's seat is the traverse crank, with the elevation crank above it.

The FuMo 214 "Würzburg Giant" radar device.